Repairing and Extending

Nonstructural Metals

VAN NOSTRAND REINHOLD'S
BUILDING RENOVATION AND RESTORATION SERIES

Repairing and Extending Weather Barriers   ISBN 0-442-20611-9

Repairing and Extending Finishes, Part I   ISBN 0-442-20612-7

Repairing and Extending Finishes, Part II   ISBN 0-442-20613-5

Repairing and Extending Nonstructural Metals   ISBN 0-442-20615-1

Repairing and Extending Doors and Windows   ISBN 0-442-20618-6

Repairing, Extending, and Cleaning Brick and Block   ISBN 0-442-20619-4

Repairing, Extending, and Cleaning Stone   ISBN 0-442-20620-8

Repairing and Extending Wood   ISBN 0-442-20621-6

Building Renovation and Restoration Series

Repairing and Extending

Nonstructural Metals

H. Leslie Simmons, AIA, CSI

VNR VAN NOSTRAND REINHOLD
New York

Copyright © 1990 by H. Leslie Simmons

Library of Congress Catalog Card Number 89-24998
ISBN 0-442-20615-1

Printed in the United States of America

Van Nostrand Reinhold
115 Fifth Avenue
New York, New York 10003

Van Nostrand Reinhold International Company Limited
11 New Fetter Lane
London EC4P 4EE, England

Van Nostrand Reinhold
480 La Trobe Street
Melbourne, Victoria 3000, Australia

Nelson Canada
1120 Birchmount Road
Scarborough, Ontario M1K 5G4, Canada

16 15 14 13 12 11 10 9 8 7 6 5 4 3 2 1

Library of Congress Cataloging-in-Publication Data

Simmons, H. Leslie.
    Repairing and extending nonstructural metals / H. Leslie Simmons.
        p.   cm. — (Building renovation and restoration series)
    Includes bibliographical references.
    ISBN 0-442-20615-1
    1. Architectural metal-work. 2. Buildings—Repair and reconstruction. I. Title. II. Series.
    TH1651.S55   1990
    693′ .7—dc20   89-24998
                                                                    CIP

Contents

CHAPTER 1   Introduction   1

CHAPTER 2 Support Systems   22

CHAPTER 3 Metal Materials   63

CHAPTER 4 Mechanical, Chemical, and Inorganic Metal Finishes   103

CHAPTER 5   Organic Coatings for Metals   155

CHAPTER 6   Design and Fabrication of Nonstructural Metal Items   225

CHAPTER 7   Installation, Repair, and Cleaning of Nonstructural Metal Items   267

Series Foreword

To spite a national trend toward renovation, restoration, and remodeling, construction products producers and their associations are not universally eager to publish recommendations for repairing or extending existing materials. There are two major reasons. First, there are several possible applications of most building materials; and there is an even larger number of different problems that can occur after products are installed in a building. Thus, it is difficult to produce recommendations that cover every eventuality.

Second, it is not always in a building construction product producer's best interest to publish data that will help building owners repair their product. Producers, whose income derives from selling new products, do not necessarily applaud when their associations spend their money telling architects and building owners how to avoid buying their products.

Finally, in the *Building Renovation and Restoration Series* we have a reference that recognizes that problems frequently occur with materials used in building projects. In this book and in the other books in this series,

Simmons goes beyond the promotional hyperbole found in most product literature and explains how to identify common problems. He then offers informed "inside" recommendations on how to deal with each of the problems. Each chapter covers certain materials, or family of materials, in a way that can be understood by building owners and managers, as well as construction and design professionals.

Most people involved in designing, financing, constructing, owning, managing, and maintaining today's "high tech" buildings have limited knowledge on how all of the many materials go together to form the building and how they should look and perform. Everyone relies on specialists, who may have varying degrees of expertise, for building and installing the many individual components that make up completed buildings. Problems frequently arise when components and materials are not installed properly and often occur when substrate or supporting materials are not installed correctly.

When problems occur, even the specialists may not know why they are happening. Or they may not be willing to admit responsibility for problems. Such problems can stem from improper designer selection, defective or substandard installation, lack of understanding on use, or incorrect maintenance procedures. Armed with necessary "inside" information, one can identify causes of problems and make assessments of their extent. Only after the causes are identified can one determine how to correct the problem.

Up until now that "inside" information generally was not available to those faced with these problems. In this book, materials are described according to types and uses and how they are supposed to be installed or applied. Materials and installation or application failures and problems are identified and listed, then described in straightforward, understandable language supplemented by charts, graphs, photographs, and line drawings. Solutions ranging from proper cleaning and other maintenance and remedial repair to complete removal and replacement are recommended with cross-references to given problems.

Of further value are sections on where to get more information from such sources as manufacturers, standards setting bodies, government agencies, periodicals, and books. There are also national and regional trade and professional associations representing almost every building and finish material, most of which make available reliable, unbiased information on proper use and installation of their respective materials and products. Some associations even offer information on recognizing and solving problems for their products and materials. Names, addresses, and telephone numbers are included, along with each association's major publications. In addition, knowledgeable, independent consultants who specialize in resolving problems relating to certain materials are recognized. Where names were not available for publication, most associations can furnish names of qualified persons who can assist in resolving problems related to their products.

It is wished that one would never be faced with any problems with new buildings and even older ones. However, reality being what it is, this book, as do the others in the series, offers a guide so you can identify problems and find solutions. And it provides references for sources of more information when problems go beyond the scope of the book.

Jess McIlvain, AIA, CCS, CSI
Consulting Architect
Bethesda, Maryland

Preface

Architects working on projects where existing construction plays a part spend countless hours eliciting data from materials producers and installers relating to cleaning, repairing, and extending existing building materials and products and for installing new materials and products over existing materials. The producers and installers know much of the needed information and generally give it up readily when asked, but they often do not include such information in their standard literature packages. As a result, there has been a long-standing need for source documents that would include the industry's recommendations for repairing, maintaining, and extending existing materials and for installing new materials over existing ones. This book is one of a series called the *Building Renovation and Restoration Series* that was conceived to answer that need.

In the thirty-plus years I have worked as an architect, and especially since 1975, when I began my practice as a specifications consultant, I have often wondered why there are so few comprehensive sources of data to help architects, engineers, general contractors, and building owners deal with the many types of existing building materials. It is often necessary to consult several sources to resolve even apparently simple problems, partly because authoritative sources do not agree on many subjects. The time it takes to do all the necessary research is enormous.

I have done much of that kind of research myself over the years. This book includes the fruits of my earlier research, augmented by many additional hours of recent searching to make this book as broad as possible. I have included as many of the industry's recommendations about working with existing nonstructural metal items as I could fit in. The data in this book, as is true for that in the other books in the series as well, come from published recommendations of producers and their associations, applicable codes and standards, federal agency guides and requirements, contractors who actually do such work in the field, the experiences of other architects and their consultants, and from the author's own experiences. Of course, no single book could possibly contain all known data about these subjects or discuss every potential problem that could occur. Where data are too voluminous to include in the text, references are given to help the reader find additional information from knowledgeable sources. Some sources of data about historic preservation are also listed.

This book, as do the others in the series, explains in practical, understandable narrative, supported by line drawings and photographs, how to extend, clean, repair, refinish, restore, and protect the existing materials that are the subject of this book, and discusses how to install the materials covered in the book over existing materials.

All the books in the series are written for building owners; architects; federal and local government agencies; building contractors; university, professional, and public libraries; members of groups and associations interested in preservation; and everyone who is responsible for maintaining, cleaning, or repairing existing building construction materials. The titles in this series are not how-to books meant to compete with publications such as *The Old-House Journal* or the books and tapes generated by the producers of the television series "This Old House."

I hope that, if this book doesn't directly solve your current problem, it will lead you to a source that will.

What This Book Contains

This book includes descriptions of most of the more commonly used metals usually found in nonstructural metal items in buildings, and the finishes that are usually applied to those metals. It also discusses common methods of fabricating, finishing, and installing nonstructural items made from metals, including the most often used methods of anchoring, fastening, and joining metals. It includes many of the industry's recommendations about repairing nonstructural metals and their finishes, as well as items made from them, and for installing those materials and items in existing spaces.

Many of the nonstructural items made from metals to which the principles discussed in this book definitely apply are listed near the beginning of Chapter 6 under the heading "Some Types of Items." Nonstructural metal items to which the principles in this book may or may not apply are listed near the beginning of Chapter 1 under the heading "What Nonstructural Metal Items Are and What They Do."

How to Use This Book

This book is divided into seven chapters that discuss the subject areas suggested by their titles. Chapter 1 is a general introduction to the subject and offers suggestions as to how a building owner, architect, engineer, or general building contractor might approach solving problems associated with dealing with existing building materials. It offers advice about seeking expert assistance when necessary and suggests the types of people and organizations that might be able to help.

Each of the other six chapters include:

- A statement of the nature and purpose of the chapter.
- A discussion of materials commonly used to produce support systems for nonstructural metal items (Chapter 2), the types of metals used in nonstructural metal items (Chapter 3), the finishes usually used on those metals (Chapters 4 and 5), methods of fabricating nonstructural metal items (Chapter 6), and methods of installing, repairing, and cleaning nonstructural metal items (Chapter 7).
- A discussion of how and why metal materials (Chapter 3), finishes (Chapters 4 and 5), or nonstructural items made from metal (Chapters 6 and 7) might fail.
- Chapters 4, 5, and 7 include lists of the types of failures that may occur in nonstructural metal items and their materials and finishes, and cross-references to the headings in those and other chapters under which possible causes of the listed failure are discussed. Thus, a failure that is recognized in the field (for example, light-colored bands across a metal surface, which is an early sign of metal fracture) can be traced from the location of that failure type in Chapter 7 by cross-reference there to several possible causes, including, among many others in this example, "Errors That Lead to Fracture" in Chapter 3. Where appropriate, there is also explanation and discussion of the

causes of the failure, so that a reader can see specific examples showing the effects of the error.

- A discussion of methods of repairing metal materials (Chapter 3) and
  their finishes (Chapters 4 and 5) and nonstructural items made from
  metal (Chapters 6 and 7) that have failed, and for installing nonstructural metal items in existing spaces (Chapter 7).
- An indication of sources of additional information about the subjects in
  that chapter.

This book has an Appendix that contains a list of sources of additional
data. These sources include manufacturers, trade and professional associations, standards-setting bodies, government agencies, periodicals, book
publishers, and others having knowledge of methods for restoring building
materials. The list includes names, addresses, and telephone numbers.
Sources from which data related to historic preservation may be obtained
are identified with a boldface **HP**. This book also has a Glossary that
includes many of the more uncommon words and usages in it.

The items listed in the Bibliography are annotated to show the book
chapters to which they apply. Entries that are related to historic preservation are identified with a boldface **HP**. Many of the publications and
publishers of entries in the Bibliography are listed in the Appendix.

Building owners, engineers, architects, and general building contractors
will, in most cases, each use this book in a somewhat different way. The
following suggestions give an indication of what some of these differences
might be.

Owners

It is probably safe to assume that a building owner who is consulting this
book is doing so because the owner's building has experienced or is now
experiencing failure of one of the materials, finishes, or nonstructural metal
items discussed here. If the problem is an emergency, the owner should
turn immediately to Chapter 1 and read the parts there entitled "What to
Do in an Emergency" and, under "Help for Building Owners," the one
called "Emergencies."

When the failure is temporarily under control, a more systematic approach is suggested. An owner may tend to want to turn directly to the
chapter containing information about a nonstructural metal item that seems
to have failed. An owner who has good knowledge of such problems, and
experience with them, may be able to approach the problem in that manner.
Otherwise, it is better first to read and become familiar with the contents
of Chapter 1, including those parts that do not at first seem to be applicable
to the immediate problem. The cause of nonstructural metal failure is not

always readily apparent, and jumping to an incorrect conclusion can be costly.

After reading Chapter 1, the owner should turn to the chapter covering the material or item that has failed. At some point, that chapter may refer to another chapter. Chapters 6, 7, and 8, for example, often refer to Chapters 2, 3, 4, and 5. When this happens it is important also to read the cross-referenced material.

A word of caution, however. Unless an owner is experienced in dealing with such failures, he or she should not simply reach for the telephone to call for professional help until after reading the chapter covering the failed material or item and applicable cross-referenced material. It is always better to know as much as possible about a problem before asking for help.

Architects and Engineers

An architect's or engineer's approach will depend somewhat on the professional's relationship with the owner. For example, an architect who has been consulted regarding a finish failure may approach the problem differently depending on whether the architect was the existing building's architect of record, especially if the failure occurred within the normal expected life of the failed finish. In such cases there may be legal as well as technical considerations. This book is limited to a discussion of technical problems.

An architect's or engineer's first impulse may be to rush to the site to determine the exact nature of the problem. For one who has extensive experience with finish failures, that approach may be reasonable. Someone with little such experience, however, should do some homework before submitting to queries by a client or potential client.

The homework might consist of reading the chapters of this book that deal with the apparent problem and consulting the sources of additional information recommended there. Then, if the problem is even slightly beyond the architect's or engineer's expertise, the next step is to read Chapter 1 and decide whether outside professional help is needed. An architect or engineer who has some related experience might delay making that decision until after studying the problem in the field. One who knows little about the subject will, however, probably want someone knowledgeable to accompany him or her on the first site visit. Chapter 1 offers suggestions about how to go about making that decision.

An architect's or engineer's approach will be slightly different when they are commissioned to renovate an existing building. In this case extensive examination of existing construction documents and field conditions is called for. When nonstructural metal failures have contributed significantly to the reasons for the renovation and the architect or engineer is not thoroughly versed in dealing with such conditions, it is reasonable to consider seeking professional assistance throughout the design process. In that event the architect or engineer should read Chapter 1 first, then refer to other chapters as needed during the design and document-production process. Even when there is a consultant on the team, the architect or engineer should have enough knowledge to understand what the consultant is advising and to know what to expect of the consultant.

Building Contractors

How a building contractor uses this book will depend on which hat the contractor is wearing at the time and the contractor's expertise in dealing with existing building materials. For the contractor's own buildings the suggestions given above for owners apply, except that the contractor will probably have more experience with such problems than many owners do.

When asked by a building owner to repair a failed nonstructural material or item, a contractor's approach might be similar to that described above for architects and engineers.

When repairs to a nonstructural metal material or item are part of a project for which a contractor is the general contractor of record, the problem is one of supervising the subcontractor who will actually repair the failed metal material or item. Even a knowledgeable contractor will sometimes find it helpful to double-check the methods and materials a subcontractor proposes against the recommendations of an authoritative source, such as those listed in this book. It is also sometimes useful to verify a misgiving the contractor might have about specified materials or methods. In each of these cases, the contractor should read the chapter covering the subject at hand and check the other resources listed whenever a question arises with which the contractor is not thoroughly familiar. Before selecting a subcontractor for repair work, a contractor might want to review Chapter 1.

Even a contractor who has extensive experience in repairing failed metal materials and items fabricated from metals will frequently encounter unusual conditions. Then the contractor should turn to the list of sources of additional information at the end of the appropriate chapter and the Appendix and Bibliography to discover who to ask for advice.

Disclaimer

The information in this book was derived from data published by trade associations, standards-setting organizations, manufacturers, and government organizations, and statements made to the author by their representatives; from interviews with consultants, architects, and building contractors; and from related books and periodicals. The author and publisher have exercised their best judgment in selecting data to be presented, have reported the recommendations of the sources consulted in good faith, and have made every reasonable effort to make the data presented accurate and authoritative. But neither the author nor the publisher warrant the accuracy or completeness of the data nor assume liability for its fitness for any particular purpose. Users bear the responsibility to apply their professional knowledge and experience to the use of data contained in this book, to consult the original sources of the data, to obtain additional information as needed, and to seek expert advice when appropriate.

Manufacturers and their products may occasionally be mentioned in this book. Such mention is intended to indicate the availability of the products and the existence of the manufacturers. No mention in this book implies any endorsement by the author or the publisher of the mentioned manufacturer or product, other products of the mentioned manufacturer, or any statement made by the mentioned manufacturer or associated in any way with the product, its accompanying literature, or in advertising copy.

Similarly, handbooks and other literature produced by various manufacturers and associations are mentioned. Such mention does not imply that the item mentioned is the only one of its kind available, or even the best available material. The author and publisher expect the reader to seek out other manufacturers and appropriate associations to ascertain whether they have similar literature or will make similar data available.

Acknowledgments

A book of this kind requires the help of many people to make it valid and complete. I would like to acknowledge the manufacturers, producers' associations, standards-setting bodies, and other organizations and individuals whose product literature, recommendations, studies, reports, and advice helped make this book more complete and accurate than it would otherwise have been.

At the risk of offending the many others who helped, I would like to single out the following people who were particularly helpful:

Henry N. Carrera, CSI, Benjamin Moore and Company, Colonial Heights, Virginia

John W. Harn, Harn Construction Co., Laurel, Florida

Jerry D. Howell, Painting and Decorating Contractors of America, Peoria, Illinois

R. A. McDermott, CSI, Benjamin Moore and Company, Montvale, New Jersey

Robert B. Molseed, AIA, CSI, CCS, Professional Systems Division, AIA Service Corporation, Washington, D.C.

Vincent R. Sandusky, Painting and Decorating Contractors of America, Falls Church, Virginia

Cathy Sedgewick, Association of the Wall and Ceiling Industries, International, Washington, D.C.

Sally Sims, Librarian, National Trust for Historic Preservation Library and University of Maryland Architectural Library, College Park, Maryland

Everett G. Spurling, Jr., FAIA, FCSI, Bethesda, Maryland

All photographs and drawings are by the author.

CHAPTER

1

Introduction

The nonstructural metal items used in a building are affected by many factors, including the construction of the building; the humidity and temperature during and after installation of the metal items; how the substrate and framing or furring to receive the metal items are installed and prepared; the metal items themselves, including the materials from which they are made, their finishes, design, fabrication, and installation; and how well the metal items are protected after they have been installed (Fig. 1-1). For interior nonstructural metal items to remain in good condition, the building's shell and those elements such as doors and windows used to close openings in the shell must turn away wind and water and protect interior spaces from excessive temperature and humidity levels and fluctuations. For exterior nonstructural metal items to resist failure they must be designed, and their materials and finishes must be selected and installed, to resist the effects of the weather and exclude water penetration (Fig. 1-2).

This chapter includes a brief general discussion of nonstructural metal items and their failures and outlines the steps a building owner can take to solve problems associated with their failure. It also discusses the rela-

1

Figure 1-1 This decorative lamp housing was located where its eventual damage was inevitable.

tionship of architects, engineers, general building contractors, specialty contractors, manufacturers, and damage, or forensic, consultants to the owner and each other on projects involving failures related to nonstructural metal items, and then it outlines orderly ways in which these professionals can approach the problem-solving process.

This chapter includes an approach to determining the nature and extent of nonstructural metal failures and suggests the type of assistance a building owner, architect, engineer, or contractor might seek to help solve a problem.

Chapters 2 through 7 contain detailed remedies and sources of additional data.

What Nonstructural Metal Items Are and What They Do

The term "nonstructural metal items" in this book refers to all metal items used in buildings, except those that are a part of the building's primary

Figure 1-2 The black material is bituminous goop slopped on in an attempt to prevent water from entering a poorly designed joint system in this metal-paneled wall.

structural foundation, framing, wall, floor, or roof support systems. A nonstructural metal item may actually be structural in nature and serve a structural function within the item itself. It may even support other construction. The difference between structural and nonstructural metals items is that nonstructural metal items are attached to and supported by the building's primary structural framing, solid substrates, or floor, wall, or roof construction. Thus, such items as loose plates, beams, and lintels, door-jamb reinforcements, steel stairs, equipment supports, and floor-opening edge-angles are here considered nonstructural, as are the other items the industry calls "metal fabrications" or "miscellaneous metals."

This book discusses the materials found in many common nonstructural metal items used in building and related construction, and their finishes. It also discusses the design, fabrication, and installation of nonstructural metal items in general. It covers the kinds of problems that can occur with such materials, finishes, and items, and indicates the industry's recommendations for dealing with many of those failures. The principles discussed in it apply to many nonstructural metal items used in buildings. Some of the items to which the principles discussed here apply are listed at the beginning of Chapter 6, under the heading "Some Types of Items."

There are many other nonstructural metal items used in buildings, however, to which the principles discussed in this book may or may not apply. At any rate, there has been no effort here to address the particular problems they may entail. The following is a listing of some types of nonstructural metal items to which the data in this book may or may not apply.

Metal flashings, gravel stops, fascias, and trim and nonstructural metal roofing, which are discussed in the book in this series entitled *Repairing and Extending Weather Barriers.*

Metal lath, furring, framing, and suspension systems associated with finishes such as plaster, gypsum board, ceramic tile, and acoustical ceilings, which are discussed in the books in this series entitled *Repairing and Extending Finishes: Part I* and *Repairing and Extending Finishes: Part II.*

Hollow steel doors and frames; aluminum, bronze, stainless steel, and other ornamental metal doors and frames; windows, glazed curtain walls, entrances, and storefronts, and associated metal louvers, finish hardware, and other door and window equipment, which are discussed in the book in this series entitled *Repairing and Extending Doors and Windows.*

Not included are prefabricated special-purpose rooms; elevators, escalators, lifts, walkways, and other transportation devices; elevated access floors; cranes; hoists; and specialized equipment items such as stage lifts and rigging, vaults, scales, barber chairs, parking gates, incinerators, and the like. However, the data here may be applicable to some of the types of metal materials and panels used in them.

Mechanical and electrical devices and equipment.

Concrete, steel, and wood structural systems and wood and metal framing and furring are discussed in Chapter 2 to the extent that failures in them affect or cause failures in nonstructural metal items. Metal partition and wall framing are discussed in detail, as is wood framing and furring. However, the design, materials, and methods used to build or repair concrete or structural metal framing, metal joists, metal decking, or similar items that form the primary structural systems, floors, or roofs of buildings are not discussed extensively. Structural metals used in pre-engineered buildings are also not specifically addressed, but some of the principles discussed here apply to some of the types of panels used in them. Also omitted are structural metal roofing and the metal bars and mesh used to reinforce concrete.

The industry divides nonstructural metal items into several loosely defined categories, but it is far from agreement about the items that fit into these categories. For example, some members of the industry call a certain class of items "miscellaneous metals," but others call that same class "metal fabrications." In a sense, all nonstructural metal items and devices that are used in buildings and related construction are fabricated from metal and are therefore metal fabrications, regardless of whether they are fabricated from sheet metals or bars, rods, plates, tubes, pipes, or shapes such as angles and T's, and no matter what metal material is used. Castings can also be called metal fabrications. The industry as a whole does not, however, accept this logic. For example, manufactured metal casework items, which are by definition fabricated from metal and are thus metal fabrications, are usually called furnishings, which they clearly are not, since they are almost always fastened securely—and permanently—in place. Fences (Fig. 1-3) are called site improvements. To avoid confusion, throughout this book all nonstructural metal items are called either "nonstructural metal items" or "metal fabrications," regardless of the names various sectors of the construction industry may have chosen to give them. Also, for the sake of convenience the two terms are used interchangeably. The author rec-

Figure 1-3 While certainly a site improvement, this beautiful old fence is surely also a metal fabrication.

ognizes that, while many of the specific items addressed will be generally accepted as belonging in the category called "metal fabrications," others will be considered by some people as misplaced. The author wishes the reader to recognize that the terms used here are for convenience only and that there is no intent to cause a new controversy, or enter an existing one, regarding which items should be placed in which category.

Failure Types and Conditions

A failed nonstructural metal item is sometimes a symptom of an underlying problem rather than a failure of the nonstructural metal item itself (Fig. 1-4). Sometimes the failure is a result of a natural phenomenon (Fig. 1-5). Recognizing that a failure has occurred often requires no special expertise. Sagging supported equipment, cracked masonry over a too-weak lintel, broken welds, and peeling finishes, for example, are easy to see, as are the kinds of damage shown in Figure 1-5. Discovering the cause of the failure and determining the proper remedy, however, often requires detailed

Figure 1-4 The falling metal trim in this photograph is a result of deterioration of the wood structure supporting it.

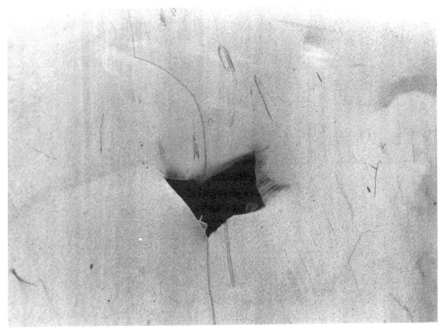

Figure 1-5   A tornado drove a piece of 2 by 4 through this sheet-steel building wall.

knowledge about the behavior of metals and their finishes and fastenings, and often of structural design as well.

When a nonstructural metal item fails, reviewing the appropriate chapter of this book, or the other books in this series where the failed material or item is discussed, might help an owner, architect, engineer, or building contractor identify the cause of the failure and solve the problem. When, after reading the material presented in this book or others in this series and examining the referenced additional data the reader does not feel competent to proceed alone, seeking professional help is in order.

What to Do in an Emergency

In an emergency, it is often necessary to act first and analyze later. When action must be taken immediately to stop damage that is already occurring or is imminent, whatever is necessary should be done. Emergency action, for example, might consist of placing a temporary support beneath a sagging piece of supported medical equipment.

A word of caution is in order here, however. Unless doing so is absolutely unavoidable, no irreversible remedial action should be taken. Small temporary repairs that cannot easily be removed can become major cost items when permanent repairs are later attempted. Hastily tearing holes in a ceiling to gain access to a failing metal fabrication support could damage the ceiling to such an extent that it must be completely removed and a new ceiling installed later.

Professional Help

If, after reading the chapters of this book or the other books in this series that address the problem area identified the next step is still unclear, or it is not possible to be sure of the nature of the problem, consult with someone experienced in dealing with the kind of failure at hand. Who that should be will vary with the knowledge and experience of the person seeking the help.

Sometimes, obtaining more than one outside opinion is desirable. Certainly this is the case when doubt exists about the first opinion obtained. For those who have little experience in dealing with failed metals a second opinion is almost always desirable.

Help for Building Owners

Building owners can turn to several types of expert help when nonstructural metals problems appear.

Emergencies. In emergencies, a building owner should seek help from the most readily available professional person or organization that can stop the damage from continuing. But even in an emergency it is better to seek help from a known organization than simply to thumb through the telephone book.

The first step is to call someone who can remedy the immediate problem—worry about protocol later. An exception to this rule should be made, however, if the building is still under a construction warranty. Then the owner should go to the person who is responsible for the warranty, who is usually a general contractor but may be a specialty contractor.

When the building is not under warranty, if the owner has a working relationship with a building construction professional able to solve the immediate problem that is the place to start. An owner who has relationships with several building professionals should start with the most appropriate one. For example, if the failed finish was recently installed by a specialty contractor under direct contract with the owner—that is, there was no general contractor involved—the owner should call that specialty contractor.

Unless an architect or engineer is the only building professional the owner knows, or the owner is unable to find someone to stop damage from continuing, calling one of them in an emergency is probably not appropriate. The best that most of them will likely be able to do is to recommend an organization that might be able to stop the damage from continuing. Thus, going through these professionals could waste valuable time.

In some areas of the country a call to a local organization representing building industry professionals will net a list of acceptable organizations specializing in repairing the kind of damage that is occurring. Such organizations might include a chapter of the Associated General Contractors of America, an association of metal fabricators, or the American Institute of Architects. Local chapters can be found by calling the affiliated national organization. Most of the applicable ones are mentioned in the appropriate chapters of this book. The names, addresses, and telephone numbers of the appropriate national organizations are listed in the Appendix.

It might be possible to find a local damage consultant who has some knowledge of the kind of metal that has been damaged, but that person may be able only to recommend someone to repair the damage.

In each case, methods that would be used under any other circumstance for contacting the various entities would apply in an emergency. There might not, however, be time to do the ordinary checking of references.

Architects and Engineers. Even in nonemergency situations, building owners with easily identifiable nonstructural metals problems often do not need to consult an architect or engineer. An exception occurs, though, when the repairing of a failed nonstructural metal item is part of a general remodeling, renovation, or restoration project. Even when the needed repairs are evidenced solely by a failure of the nonstructural metal, repairs requiring the manipulation of structural systems or that might cause harm to the public may require a building permit. The repair of damage to some metal fabrications themselves, as stairs, for example, may also require a government permit. In many jurisdictions, building officials will not consider permit applications unless they are accompanied by drawings and specifications signed and sealed by an architect or engineer. Sometimes the scope of a problem warrants hiring an architect or engineer to prepare drawings and specifications even when the authorities having jurisdiction do not require these documents. Design considerations will occasionally make hiring an architect desirable whether or not doing so is required by law.

When an owner hires an architect or engineer, the same standards that one would usually follow when hiring such professionals apply. In addition to their having the usual professional qualifications applicable to any project, however, architects and engineers commissioned to perform professional services related to an existing building should have experience in the type of work needed. It is seldom a good idea, for instance, to hire an architectural firm with experience solely in office-building construction, no matter the scope, to renovate an existing hospital.

The expected life span for many nonstructural metal items may be the same as that of the building they are a part of, but nothing can be expected to remain in a pristine state forever. Therefore, when a nonstructural metal item that has not reached its expected life span fails in a building that was designed for the owner by an architect, the owner should consider whether to inform the architect and request assistance in solving the problem. In some cases the architect will have a legal or contractual obligation to help. Discussion of the architect's potential legal or contractual obligations in such cases is beyond the scope of this book. Most architects will be interested enough to become actively involved, however, even when they are not required to do so by law or contract provision. The architect may help determine the cause of the failure and suggest solutions. He or she may also help the owner determine which other professionals, fabricators, or manufacturers to contact and will critique their recommendations. Depending on the circumstances, reimbursement for the architect's services may or may not be appropriate. Unless the owner has a long-term relationship with, or is currently a client of, the architectural firm, a frivolous complaint is likely to result in a bill for services rendered. For example, paint peeling from a fifteen-year-old metal fabrication is probably not the architect's responsibility under the original agreement between owner and architect. On the other hand, an architect should not expect payment for work related to legitimate complaints about failures caused by the architect's negligence, ignorance, or incompetence.

General Building Contractors.   Nonstructural metal failure problems that do not involve the building's structure or design are often best handled directly by a building contractor. Depending on the scope of the problem, the contractor chosen may be a general contractor or a specialty contractor. When the problem is an isolated one that involves a single discipline, a single specialty contractor may be all that is needed. When multiple disciplines are involved, however, a general contractor may be needed to coordinate the activities of several specialty contractors. Owners with experience in administering contracts may be able to coordinate the work themselves without employing a general contractor.

When selecting a general contractor it is best to select one the owner knows. If there is no such contractor, the owner should then seek advice in finding a competent firm. Such advice might come from satisfied building owners, architects, or engineers the owner knows, or through the American Institute of Architects. As a last resort an owner might ask for a recommendation from a contractors' organization. The owner should bear in mind, however, that asking for a recommendation from an organization that is supported by and represents the firm being sought will, at best, result in a list of firms in the area. This is not a good way to get a recommendation. To say this is in no way to make an indictment of contractors or their associations. The same advice applies to locating reputable doctors, lawyers, and architects—you find the best ones by word of mouth.

If a project is large enough to warrant doing so, an owner might consider seeking competitive bids from a list of contractors. Few nonstructural metal-repair projects are that large or complicated, however. Negotiation is better, if reputable firms can be found with which to negotiate. If all parties are unknown, competitive bidding may be the only way to get a reasonable price.

When a nonstructural metal item fails in a building that is still under a construction warranty, or the failed metal is still under a special warranty, the owner should contact the building contractor responsible for the warranty. The contractor's legal and contractual obligations under warranties may be complex and difficult to determine in specific cases. They are at any rate beyond the scope of this book. Under ordinary circumstances an owner should not have to make the initial contact with the contractor's subcontractor or supplier for the failed finish unless the general contractor cannot, or will not, do so. The owner should contact the architect and his or her own attorney for guidance about warranties and legal matters.

Even after the general and special warranties have expired, when a nonstructural metal item that has not been installed for a long (this period is subjective) time fails, the owner should probably still contact the general contractor who built the building and request assistance in solving the problem. The owner might consider calling another contractor if dissatisfied with the original contractor for some reason, but doing so may be costly. A new contractor will not be familiar with the building's peculiarities.

Fabricators and Manufacturers. Some nonstructural metal items are made up of manufactured components that are shipped to the construction site and assembled there in place on, or in, the building. Others are fabricated specifically for the project in a shop, using metal sheets, plates, and shapes sometimes combined with manufactured products. These units are then shipped to the construction site either completely assembled or partly disassembled to permit shipment.

Although some such firms are more helpful than others, the fabricator or manufacturer of an item that has failed or an association representing the manufacturers or fabricators of such products are often knowledgeable sources of recommendations for dealing with nonstructural metal item failures. They may be the best qualified agencies to determine the cause of a failure.

In very simple cases, asking the manufacturer for advice may not be necessary. In more complex cases, however, contractors and architects will often bring in the failed nonstructural metal item's fabricator or manufacturer to help determine the cause of, and solution to, a failure. When they do not do so, the owner probably should. When the failure is extensive or solving it will be expensive or highly disruptive to the owner's activities, the owner may want to bring in a second manufacturer or fabricator who makes a product similar to the failed one, to verify the advice given by the first manufacturer or fabricator. Knowledgeable and reputable people often do not agree about causes of failures and the methods needed to repair them and prevent future failures. In some cases the material itself may have failed, or the manufacturer's or fabricator's installation instructions may have been faulty. Then the owner might be justified in not agreeing to throw good money after bad by dealing further with that manufacturer's or fabricator's products or advice. A new product or method may be needed to solve the problem.

Some fabricators and manufacturers install the items they make; others do not. When the fabricator or manufacturer does not install its own product, refer to the discussion in the following section entitled "Specialty Contractors."

When a repair needed to a nonstructural metal item is small and simple and there is no general contractor or related specialty contractor involved in the project, it is often best for a building owner to contact first the fabricator or manufacturer, as applicable, of the damaged item. Even those manufacturers and fabricators who do not, or cannot, install or repair a damaged item themselves can often help an owner find someone to repair a failed nonstructural metal item. An owner who does not wish to deal with the original fabricator or manufacturer, or does not know who they are, can use the methods suggested earlier for finding a general contractor to locate a reputable fabricator or manufacturer. In addition, general contractors that the owner knows to be reputable can be a source of recommendations for manufacturers and fabricators, as long as there is no symbiotic relationship between the general contractor and these parties. Specialty contractors who install items similar to those to be repaired are also excellent sources of the names of fabricators and manufacturers.

Competitive bidding is usually an acceptable procedure when dealing with fabricators on small projects, but some may not be interested if the project is too small or narrowly specialized.

Product manufacturers may or may not agree to a competitive bidding procedure. Most will not be interested in manufacturing small quantities of a product that they do not usually make. Therefore, when a failed metal item is a manufactured product, it may be necessary to deal with the original manufacturer, regardless of the owner's wish not to do so.

When a failed nonstructural metal item is under warranty, the fabricator or manufacturer who made the item will probably be required to make needed repairs or provide a replacement item, but the owner should not contact the fabricator or manufacturer directly if there was a general contractor involved in the original project. The general contractor, who is usually responsible for the warranty, may choose to have repairs made by, or a replacement item provided by, an entirely different subcontractor. There is normally no direct contractual relationship between the owner and the fabricator or manufacturer in such a case.

When there is no general contractor or specialty contractor involved, the owner should contact the manufacturer or fabricator, who is usually directly responsible for any warranty that may exist.

When a nonstructural metal item has failed because it was improperly fabricated or installed and it is no longer under warranty, the owner may want to consider using a different fabricator or manufacturer to effect repairs or to fabricate and install a replacement item.

Specialty Contractors.   As mentioned earlier, some fabricators and manufacturers of nonstructural metal items install the items they make, but others do not. When the fabricator or manufacturer does not install its product, sometimes a general contractor will install the item using his or her own forces and will sometimes employ a specialty contractor to install it.

When a repair needed to a nonstructural metal item is small and simple, if there is no general contractor on the project and the fabricator or manufacturer of the item does not install its own product, it may be best for a building owner to deal directly with a specialty contractor. The methods suggested earlier for finding a general contractor may also be used to find a reputable specialty contractor. In addition, general contractors the owner knows to be reputable are a source of recommendations for specialty contractors, as long as there is no mutually beneficial relationship between the general contractor and the specialty contractor. Firms that fabricate or manufacture the items to be repaired are also excellent sources of the names of specialty contractors, and vice versa.

Competitive bidding is an acceptable procedure when dealing with specialty contractors on small projects.

When a failed nonstructural metal item is still under warranty the specialty contractor who installed it will probably be required to make the repairs, but the owner should not contact the specialty contractor directly when there was a general contractor involved in the original project. The general contractor, who is usually responsible for the warranty, may choose to have repairs made by an entirely different subcontractor. There is normally no direct contractual relationship between the owner and the specialty contractor in this case.

When there is no general contractor, the owner should contact the specialty contractor, who is usually directly responsible for any warranty that may exist.

When a nonstructural metal item has failed because it was improperly installed and it is no longer under warranty, the owner may want to consider using a different specialty contractor to effect repairs or install a replacement item.

Specialty Consultants.   There are three sets of circumstances in which an owner may want to look for a consultant who is knowledgeable about the type of nonstructural metal failure that has occurred.

First, consider hiring a qualified consultant when a knowledgeable contractor or architect or a reputable and trustworthy manufacturer's or fabricator's representative is not available for consultation. Contractors and architects with great knowledge about relatively recent materials and their failures may nevertheless know little about the types of materials and systems used in older buildings. There may not even be a manufacturer who has knowledge of a product that is no longer manufactured. A requirement for historic preservation may be sufficient cause to look for a specialty consultant.

Second, an owner may want to employ a knowledgeable consultant when the owner's contractor and architect and the fabricator or manufacturer of a failed item cannot agree on the cause of the failure, the means appropriate to make repairs, or which party, if any, is responsible for the failure. For some failures there may be no one to blame. No one can prevent nonstructural metal items from being damaged by an earthquake, for example, or by an automobile's running into a fence (Fig. 1-6). For other failures, though, blame can often be placed (Fig. 1-7). Battles about responsibility can be long and difficult to resolve. Too frequently they end up in court. Sometimes, however, an outside "expert" can help resolve conflicts and prevent the parties from having to resort to litigation.

Third, using a consultant may be desirable when the people who are already involved cannot determine with certainty the cause of the damage or the proper method for making repairs.

Consultants who specialize in damage problems are sometimes called forensic consultants. They should be able to determine the nature of a problem and identify its true cause, find a solution to the problem, select

Figure 1-6 No reasonable design or installation procedure could have prevented this damage.

the proper products to use in making repairs, write specifications and produce drawings related to the solution, and oversee the repairs.

It may not be easy to find a damage consultant who specializes in nonstructural metal damage or metal failure. It may be somewhat easier to find a consultant with knowledge about the particular metal item that has failed. Unfortunately, all damage consultants are not created equal. Many who represent themselves as consultants are actually building product manufacturers' representatives trying to increase their sales, or specialty contractors trying to enlarge their businesses. Most of them are reputable and some are competent to give advice, but few are sufficiently knowledgeable to identify underlying substrate or structural problems or advise an owner about solving them. Following the recommendations of an incompetent consultant can cause problems that may linger for years.

The process of selecting a consultant can be filled with uncertainty and

Figure 1-7 Proper maintenance could have prevented a small damaged place from becoming the gaping hole in this metal panel.

the potential for harm. There are no licensing requirements and no nationally recognized associations representing consultants who specialize in non-structural metal damage. As a result, a building owner with nonstructural metal item problems who needs to hire a damage consultant must determine the qualifications of that consultant with little help. One way to do so is to hire an architect or engineer and let them select and qualify the consultant, subject of course to the owner's approval.

Asking a specialty contractor to recommend a consultant may not be a good idea. Although some of them hire damage consultants themselves, many specialty contractors feel that such consultants are, at best, a necessary evil. This opinion probably stems from the tendency of some manufacturers' representatives to call themselves consultants and then oversell their abilities and knowledge.

Regardless of who makes the recommendation or does the hiring, the damage consultant should have demonstrated expertise in dealing with the problems at hand. Obtaining references from the consultant's satisfied clients is an appropriate prequalification tool. A licensed architect or engineer who has extensive experience dealing with existing construction might be an acceptable damage consultant.

A consultant who is qualified to deal with nonstructural metal failures in relatively new construction may know little about the kinds of materials that might be encountered in very old buildings. This is a particular problem when historic preservation is involved. A consultant for that type of work needs to be able to demonstrate knowledge in dealing with old materials and systems and the special requirements associated with historic preservation. It may not be possible to find a consultant who is an expert in dealing with old materials who is also knowledgeable in general construction principles. In this case it may be necessary to find two consultants with complementary knowledge.

A damage consultant should have extensive knowledge of the material that has failed, in addition to an understanding of general construction principles and the physiology of structures. Many nonstructural metal failures are a result of not one but several problems. The consultant should know enough about buildings as a whole to be able to identify all of the underlying problems, not just the obvious ones or those directly associated with the metal item itself. Many nonstructural metal item problems are caused by movement of the supporting structure, for example. The consultant should be able to determine whether the movement responsible is normal movement that has not been accounted for in the metal item or is structure failure that might require repairing the structure itself. The consultant need not, however, be a structural engineer or know how to repair the structure. Usually, the ability to make a diagnosis is sufficient.

Finally, the consultant should have no financial stake in the outcome of the investigation. The owner needs to be sure that the consultant is an independent third party who is selling a professional service, not the installation or repair of a particular product. Neither a product manufacturer's representative nor a contractor who wants to make the repairs meets this qualification.

Help for Architects and Engineers

Architects and engineers usually get involved in repairing nonstructural metal items only when the failure occurred in a building they designed or when the repairs are extensive, have occurred on a prestigious building, or are part of a larger renovation, restoration, or remodeling project.

Architects without extensive experience in dealing with nonstructural metal problems or working with existing nonstructural metals and their finishes should seek outside help from someone who has such experience.

The nature of that help, and the person selected to consult on it, depends on the type and complexity of the problem.

Other Architects and Engineers.   One source of professional consultation for architects and engineers is other architects or engineers experienced with the type of nonstructural metals problem at hand. The qualifications needed for such a consultant are similar to those outlined earlier in this chapter for specialty damage consultants. The recommended architect or engineer need not be participating currently in an architectural or engineering practice. Specifications consultants and qualified architects and engineers employed by government, institutional, or private corporate organizations should not be overlooked.

Product Fabricators and Manufacturers, and Industry Standards. Fabricators, manufacturers, and their associations often provide sufficient information to enable a knowledgeable architect or engineer to deal with many nonstructural metal failures. They are often the most knowledgeable source of data about failures of their products and related corrective measures.

The architect or engineer must, of course, compare fabricators' and manufacturers' statements with those of other fabricators and manufacturers and with industry standards, then exercise good judgment in deciding which claims to believe. The problem is the same as any other in which architects and engineers consult producers and their associations for advice. The architect or engineer must study every claim carefully, especially if it seems extravagant, and double-check everything. Claims by the fabricator or manufacturer of a failed product that identify causes other than product failure should be examined especially carefully. Here a second opinion may be helpful.

Refer also to "Fabricators and Manufacturers" under "Help for Building Owners," earlier in this chapter.

Specialty Consultants.   Architects sometimes employ independent specialty damage consultants, who may or may not be architects or engineers. The qualifications outlined earlier in this chapter for such consultants remain applicable, no matter who is hiring the consultant.

Help for General Building Contractors

Whether a general building contractor will need to consult an outside expert depends on the complexity of the problem, the contractor's own experience, and whether the owner has engaged an architect or specialty consultant. Duplication of effort is unnecessary unless the contractor intends to challenge the views expressed by the owner's consultants or the content of the drawings or specifications.

A general building contractor acting alone on a project where the owner has not engaged an architect, engineer, or consultant must base the need for hiring consultants on various factors. These include the contractor's own experience and expertise with the types of problems that will be encountered, the contractor's specialty subcontractor's experience and expertise in dealing with the types of problems involved, and the complexity of the problems.

Hiring a specialty consultant may complicate a general contractor's relationship with his or her subcontractors. Such overlapping responsibility is seldom justified and is often a bad idea. An experienced, qualified specialty subcontractor is not likely to appreciate a damage consultant hired to tell the subcontractor how to do the job. A general contractor would be better off finding a qualified subcontractor to do the work and relying on that subcontractor's advice. If no such subcontractor is available and the general contractor is not experienced with the problem at hand, hiring a consultant may be necessary, regardless of the feelings of the subcontractor. Muddling along to salve feelings is wholly incompetent and unprofessional. In such circumstances the contractor should recommend that the owner employ a qualified architect or engineer to specify the repairs and let the owner and the design professional hire a damage consultant, if necessary.

If a contractor should decide to hire a specialty damage consultant, the recommendations earlier in this chapter for owners and architects apply.

It is often necessary for a general contractor working with an owner who has not hired an engineer or architect to consult with the fabricator or manufacturer of a failed nonstructural metal item. The fabricator's or manufacturer's representative can often identify the causes of failure and recommend repair methods. When the contractor has engaged a specialty subcontractor, the fabricator or manufacturer often proves a source for a second opinion about such matters. The same precautions mentioned earlier for owners and architects dealing with a fabricator or manufacturer apply when a contractor deals directly with them.

Prework On-site Examination

On-site examinations before work begins are important tools in helping to determine the type and extent of a nonstructural metal failure and the damage to underlying construction it might portend. Who should be present during an on-site examination is dictated by the stage at which the examination will take place.

The Owner

The first examination should be by the owner or the owner's personnel to determine the general extent of the problem. This examination should help the owner decide what the next step should be and the type of consultant the owner needs to contact, if any.

After the owner has selected a consultant, owner and consultant should visit the site and define the work to be done. The owner's consultant may be an architect or engineer, general contractor, specialty contractor, specialty damage consultant, or manufacturer or fabricator. This second site examination should be attended by a representative of each expert that the owner has engaged to help with the problem. If a specialty damage consultant is engaged by another of the owner's consultants, he or she should also be present. And the general contractor's specialty subcontractors should be present, if they have been selected.

During the second site visit, the parties should become familiar with conditions at the site and offer suggestions about how to solve the problem.

Architects and Engineers

An architect or engineer hired to oversee repair of a failed nonstructural metal item should, before visiting the site if possible, determine the components used in the failed item and the type of underlying and supporting construction, and examine the shop and setting drawings for the item, if they are available. The architect or engineer should then visit the site with the owner to determine the extent of the work needed and begin to decide how to solve the problem. After discussion with the owner about the nature of the problem, the architect should decide whether to engage professional help or contact the failed product's fabricator or manufacturer. If a consultant is to be used, that person should visit the site with the owner and the architect. If the fabricator or manufacturer is to be brought in, its representative should accompany the owner and the architect on the site visit.

If as a result of the architect's first site visit the architect determines that a specialty damage consultant or fabricator's or manufacturer's representative, previously considered unnecessary, is now needed, the architect should arrange for another site visit with that consultant or fabricator's or manufacturer's representative.

During the progress of the work the architect, the architect's consultant, and the owner's representative should visit the site as often as necessary to determine fully the nature and extent of the problem and help arrive at a full solution. These site visits should extend the scope of the observation beyond the immediate problem to ascertain whether additional, unseen, damage might be present.

Building Contractors

Nonbid Projects.   On nonbid projects, a building contractor may wear two or more hats.

The easiest situation to deal with is a negotiated bid based on professionally prepared construction documents. In this case the contractor should conduct an extensive site examination to verify the conditions shown and the extent and type of work called for in the construction documents. Offering a proposal based on unverified construction documents is a bad business practice that can cost much more than proper initial investigation would have, if the documents are later found to be erroneous.

When the owner has not hired an architect or consultant to ascertain and document the type and extent of the work, the contractor must act as both designer and contractor. If so, the contractor should visit the site with the owner as soon as possible, and revisit as often as necessary, to determine the nature of the problem and the extent of the work to be done. The contractor may choose to hire a specialty consultant or specialty subcontractor or both, or to consult with a representative of the fabricator or manufacturer of the failed nonstructural metal item. Those individuals should then accompany the contractor on the site visits and participate in forming the contractor's recommendations. A carefully drawn proposal is an absolute must to be sure that the owner does not expect more than the contractor proposes to do.

Even when the owner hires an architect or other consultant, the contractor should still visit the site with the owner and the owner's consultant as soon as possible. The purpose of this visit is to ascertain the extent and type of work to be done and to recommend repair methods. The contractor should also invite his or her specialty subcontractors to visit the site with the owner, the owner's consultants, the fabricator's or manufacturer's representative, and the contractor, if possible. The more input the contractor has in the design process, the better the result is likely to be.

Bid Projects.   Even when the contractor is invited to bid on a project for which construction documents have been prepared, a prebid site visit is imperative. No contractor should bid on work related to existing construction without extensive examination of the existing building. Some construction contracts may even demand it. Some contracts in fact try to make failure to discover a problem the contractor's responsibility. Even if the courts throw out that clause, who can afford the time and costs of a lawsuit? A contractor should establish exactly what work is to be done before bidding on it. Insufficient data may be cause for choosing not to bid on a project.

CHAPTER
2

Support Systems

Nonstructural metals are defined for this book as metals and items made from metals that are supported by either the building's structural framing system; solid substrates, such as concrete and unit masonry, which may be either structural or nonstructural; or framing and furring systems erected specifically to support the metal or to support the metal as well as other building components such as finishes. The failure of a nonstructural metal item may be caused by the failure of its supports or attributed to problems with other building elements, such as roof or wall construction.

This chapter includes a general discussion of steel and concrete structural systems, concrete and masonry solid substrates, and other building elements, and discusses possible failures in them that could lead to nonstructural metal items' failure.

Also included are descriptions of the types of metal and wood framing and furring that might be used to support nonstructural metal items. In each case there is a discussion about the components and materials normally used, potential errors that can lead to failure in the framing or furring or in the metal supported, and ways to correct the errors once they have been identified.

Problems in a building's structural system, solid substrates, or other building elements, or in the framing and furring systems supporting a nonstructural metal, sometimes appear first as a failure in the metal item being supported. This chapter and Chapters 3 through 7 contain discussions of such causes of failures as improper design, bad workmanship, and others. Chapters 4, 5, and 7 each contain a section called "Evidence of Failure," which lists many types of failures that might occur in metal materials, their finishes, and in nonstructural items made from metal. Some of those evidences of failure are referenced to headings in another portion of the same chapter where possible causes of that type of failure are discussed. Others are cross-referenced to headings in other chapters. For example, among the causes of failure given in Chapters 4, 5, and 7 are "Concrete and Steel Structure Failure," "Concrete and Steel Structure Movement," "Metal Framing and Furring Problems," "Wood Framing and Furring Problems," "Solid Substrate Problems," and "Other Building Element Problems." These six reasons for failure are addressed in this chapter under those headings.

Concrete and Steel Structural Framing Systems

A building's structural system can have a major impact on whether a nonstructural metal item supported by it remains in good condition or fails. The structural system can affect the nonstructural metal item in one of two ways: the structure itself can fail, or the structure can move in ways not anticipated in the nonstructural metal item's design.

As defined here, steel structural systems include steel columns, beams, girders, trusses, bar joists, floor and roof decks, and the foundations that support the structural steel. Also included are pre-engineered and prefabricated steel buildings.

Concrete structural framing systems are defined here to include both cast-in-place and precast concrete footings, columns, beams, girders, slabs, and related components like stairs. This category also includes composite construction made using both steel and concrete. Composites are used most often for beams and slabs, but they may also be used for columns and other structural components.

Concrete and Steel Structure Failure

Concrete and steel structural systems fail either because they are improperly designed or because they experience unanticipated conditions that exceed their design limitations. These limitations are dictated by materials characteristics, legal requirements, and economic factors. It is not economically feasible to design every structure to handle every condition that might occur. Even a building designed to withstand an earthquake measuring 8 on the Richter scale may fail if the level reaches 8.5. It might not be possible to design a building that would remain completely undamaged in even a small earthquake.

Structure failure may be either small or large in scope, from slight damage to complete building collapse, but most structure failure is relatively small in magnitude. A single cause may generate failure at any level. An undersized footing may, for example, lead to building collapse, or simply to more settlement than normal.

Because even minor structural failure may damage supported nonstructural metal items, it is necessary before repairing failed nonstructural metals to determine whether structural failure is responsible for a metal item's failing. Where structural failure is to blame, it is usually necessary to correct the structural failure before repairing the nonstructural metal item. Otherwise, the metal item's failure will usually recur. In severe cases structural reconstruction, such as shoring up beams, adding columns, or replacing structural members, may be necessary. When structural damage is self-limiting and not dangerous to the building or people, however, it is sometimes possible to modify the existing support system or provide a new one for the nonstructural metal item without making major corrections to the failed structure. An example is a self-limiting minor settlement caused by a small weak spot in the earth beneath a portion of a footing.

The repair of a failed concrete or steel structural system is beyond the scope of this book. It has been written with the assumption that failures in such systems have been diagnosed and necessary repairs have been made.

Structural system failure will probably be due to one of the following causes.

The structure may not have been properly designed to withstand the loads to be applied without undergoing excess deflection, vibration, settlement (especially differential settlement), expansion, or contraction.

The building's foundations may not have been properly designed or installed. The designer may have used spread footings when pilings should have been used. Maybe no borings or soil bearing tests were made at the site before the design was prepared, or the tests were improperly made. Perhaps tests were properly made, but were then either ignored or improperly interpreted by the designer. The bearing soil beneath the footings was perhaps permitted to become wet during construction, or was improperly excavated to too low an elevation, then backfilled with soil to conceal the error. There are many other possible reasons for this error, but the result is the same: excessive or uneven settlement.

A further cause of structural system failure may have been an unforeseen traumatic event that may have occurred so that forces were applied to the building in excess of those it was designed to resist. Earthquakes, hurricanes, and tornadoes are examples of such unpredictable events.

Concrete and Steel Structure Movement

In addition to failures caused by concrete or steel structure failure, damage to nonstructural metals can occur because the structure moves, especially if the movement is larger than expected. Structure movement should be suspected when a nonstructural metal fails, particularly if the evidence of the failure consists of broken or open joints or of an object fabricated from nonstructural metal's separating from adjoining construction. To prevent failure, the design, fabrication, and installation of nonstructural metal items must take structure movement into account.

Undue structure movement may be a symptom of structural failure, as just discussed. Some structure movement is, however, normal and unavoidable. The expected structural movement due to wind, thermal expansion and contraction, and deflection under loads is large in many modern buildings, which are purposely designed to have light, flexible structural frames that are less rigid than the structural systems used in most older buildings. Exterior column movement is a particular problem. Movement may be especially large in high-rise structures where both flexibility and wind loads are large. While these light modern designs are usually safe structurally, they may contribute to failures in nonstructural items made from metals, unless the designer is aware of the problems they impose and takes steps to head them off. Nonstructural items made from metals must be designed to accommodate the amount of expected movement. Normal structure movement may result from one of the causes that follow.

Variable wind pressure, particularly on high-rise structures, can cause considerable structural movement.

Structure settlement will also cause some movement. It is almost impossible to eliminate all building settlement, although proper design will keep such movement to a minimum.

All materials expand and contract when the temperature changes. Unfortunately, some materials change a great deal more than others. Differential thermal expansion and contraction is a major cause of failure in building components.

Deflection of structural members and slabs will cause some movement in buildings. Deflection is usually a temporary condition. When the loads are removed, the structural elements often return to their original shape. Permanent deflection can occur, however, when the applied loads are large.

The particular type of deflection known as creep, which occurs in concrete structures, is usually not reversible, however. Creep is the process of the structural elements of a building permanently changing their shape because of initial deflection due to their own weight.

Another cause of movement in buildings is structure vibration, which is often transferred from operating equipment in the building. Vibration can loosen fasteners.

Metal Framing and Furring

Some of the nonstructural metal items addressed in this book are often fastened to, or supported by, metal framing and furring, often through an applied finish, such as plaster or gypsum board. The metal framing discussed here includes nonbearing wall and partition framing, which consists of C-shaped and truss studs and special studs used in shaft wall systems. It also includes metal stud-type framing designed to bear loads, which is constructed of heavier-gage components and often requires different attachment devices than does nonbearing framing. The latter is commonly called "cold-formed metal framing."

The metal furring discussed here includes wall furring only. Suspended nonstructural metal items are usually attached directly to overhead structural elements.

Metal structural framing, such as steel beams and girders, and bar joists, is not included in the following discussion.

Materials

Metals and Finishes.   In exterior installations and interior ones where high humidity will be present, as in swimming pool areas or commercial kitchens, metals used as framing or furring should be hot-dip galvanized steel. Zinc-alloy metals are used to support some finishes, but they are usually not used in conjunction with the types of nonstructural metal items discussed in this book. The galvanized finish should be equal to that stipulated in ASTM Standard A 525 as G90 for 18 gage and lighter formed metal products, and ASTM Standard A 123, galvanized after fabrication, for 16 gage and heavier products.

In other locations the manufacturer's standard steel products with the manufacturer's standard galvanized finish may be used where light-gage metals are required. Where 16-gage or heavier sheet metal is required in locations other than those mentioned above, products should be rolled steel and formed sheet steel with a rust-inhibitive paint finish. Where 7 gage,

3/16 inch thick, or heavier rods and bars are required, the finish may also be rust-inhibitive paint.

Furring Members.   The metal furring members usually used to support or fasten the types of nonstructural metal items covered in this book include C-shaped channels, Z-shaped furring members, hat-shaped channels (Fig. 2-1), and studs. The studs used for furring should be the same type of studs used in nonbearing partitions.

The C-shaped channels for this purpose are usually 16-gage, hot-rolled or cold-rolled steel channels that are either 3/4 inch, 1-1/2 inches, or 2 inches deep. When painted, 3/4-inch deep channels should weigh not less than 300 pounds per 1,000 linear feet, 1-1/2 inch deep channels should not weigh less than 475 pounds per 1,000 linear feet, and 2-inch deep channels

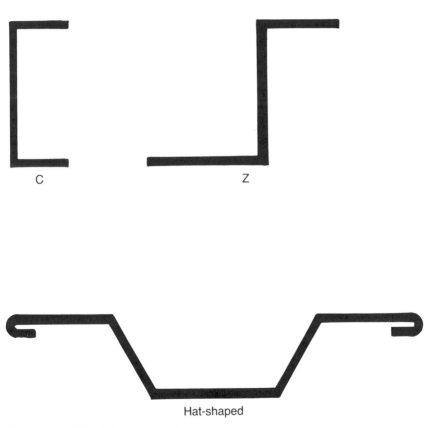

C   Z

Hat-shaped

Figure 2-1   Metal furring members.

should weigh at least 590 pounds per 1,000 linear feet. Galvanized channels will be slightly heavier.

Furring to receive screw fasteners is usually nominally 20- or 25-gage, hat-shaped channels complying with ASTM Standard C 645 (see Fig. 2-1). Most hat-shaped furring members have 1-3/8 wide by 7/8-inch deep crowns and flanges for fastening in place, but 1-1/2 inch deep, hat-shaped furring members are also used sometimes.

Some installations use Z-shaped members formed from 25-gage sheet steel (see Fig. 2-1). Their wall legs should be not less than 3/4 inch wide. The facing legs should be at least 1-1/4 inches wide. The depths of Z-shaped furring members will vary, but they are usually at least 1 inch deep.

The types of corner, resilient, and other special furring members used to support finishes are seldom required for the types of nonstructural metal items covered by this book.

Nonbearing Wall and Partition Framing.   There are three types of steel studs that might be encountered in an existing building. Each type has its own distinctive floor and ceiling runners.

Open-Web Stud Framing Systems.   There are several different open-web stud designs, including the currently popular wire-truss type. Figure 2-2 shows three designs, but others may be found in an existing project. Most open-web studs are non–load bearing.

Wire-truss type studs include both those with double wire flanges and those with steel angle flanges. The wires are usually 7 gage, angle flanges 16 gage.

Open-web studs also include those with hollow sheet-metal tube flanges designed to accept fasteners. These are sometimes called "nailable."

The webs of open-web trusses are made either of wire woven across the gaps between the flanges to form a truss or of flat sheet-metal strips welded to or clamped in place by the flanges.

Open-web studs have floor and ceiling runners specifically designed to receive them. The runners are often notched or otherwise configured to control the studs' spacing. Some open-web studs are anchored to runners by specially designed shoes. Others are held in place by wire ties, clips, screws, or locking tabs. Open-web studs are often found beneath standard thickness (3/4 inch thick or more) plaster or stucco that is applied over metal lath.

Channel Stud Framing Systems.   Channel studs are open, C-shaped sheet-metal studs (Fig. 2-3) designed primarily for screw attachment of gypsum board materials. They may also be found behind standard-thickness plaster with gypsum or metal lath, veneer plaster, stucco, and other finishes.

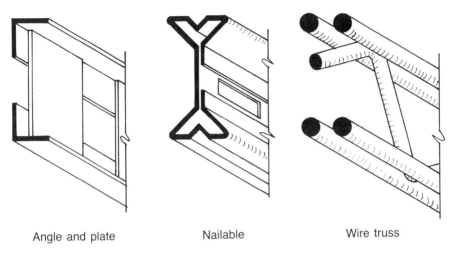

Angle and plate	Nailable	Wire truss

Figure 2-2 Open-web metal studs.

Non–load bearing channel studs should comply with ASTM Standard C 645. (Load-bearing studs are addressed later in this chapter under "Cold-formed Metal Framing.") Most non–load bearing channel studs are made from either 20, 21, 25, or 26 nominal-gage steel sheets formed into C-shaped units. The better studs have each flange returned to form a stiffening lip parallel to the web. Channel studs have knockouts or preformed holes in their webs to permit pipes and conduits to pass through.

Floor and ceiling runner channels for channel studs are C-shaped units with flanges about 1-1/4 inches high. They are designed to receive the studs.

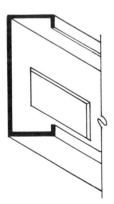

Figure 2-3 Metal C-shaped stud.

Special Stud Framing Systems. The types of special studs include specially shaped (C-H, H, etc.) ones designed for use in shaft wall systems (Fig. 2-4). The types vary with each manufacturer. Shaft wall systems also include specially shaped runners, such as J-shaped units. Special runners vary in size and weight with each shaft wall system's manufacturer.

Sheet-metal angles are also sometimes used as floor and ceiling runners for special applications such as solid gypsum board walls. Angle runners are usually formed from 24-gage metal into 1-3/8 by 7/8-inch or larger angles.

Cold-formed Metal Framing Members. Components falling into the cold-formed framing category are used for stud walls, joists, metal wall panel backup, and many other support conditions.

Studs for use in cold-formed metal framing are load-bearing in both the transverse and the axial directions. Studs and runners should both comply with ASTM Standard C 955. Cold-formed metal framing studs are usually formed from nominal 18-gage or heavier sheet steel. Incidental parts, however, such as some bracing members, may be formed from nominal 20-gage sheet steel.

Cold-formed metal framing studs may be either C-shaped or punched-channel studs. Runners are C-shaped.

Cold-formed joists and other framing components are formed from sheet metal. The component parts making up such elements are often, but not always, C-shaped.

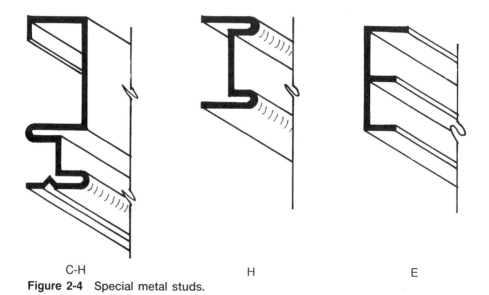

C-H   H   E

Figure 2-4 Special metal studs.

The size, thickness, and shape of cold-formed metal framing components are dictated by the type of component and the loads it must support. Framing components are usually prefabricated, often in the shop. Like members should be fastened together by being welded. Dissimilar parts may be fastened together by welding, screwing, or bolting. Components should not be joined with wire.

Framing and Furring Accessories and Other Materials. Accessories include fasteners, bolts, screws, anchor clips, wire, and other items necessary to the complete installation of framing and furring.

Attachment Devices. The types of devices used to install framing and furring include those that follow.

Tie wire should be Class 1 galvanized soft-temper steel wire that complies with ASTM Standard A 641. Ties for use with studs should be not less than 18-gage wire; for use with furring, not less than 16 gage.

Clips should be the framing or furring manufacturer's standard galvanized wire-type clips.

Fasteners should comply with ASTM Standard C 754. Screws should be the rust-inhibitive, oval pan head, self-drilling, and self-tapping types designed for use with special power-driven tools. They should comply with ASTM Standard C 646. Bolts should be suitable for the purpose.

Other Materials. Several other materials are used in framing and furring systems, including those that follow.

Asphalt-saturated felt should be 15-pound nonperforated roofing felt that complies with ASTM Standard D 226.

Acoustical sealant should be a water-based, nondrying, nonbleeding, nonstaining type that is permanently elastic.

Most sound-attenuation blankets are semirigid mineral fiber blankets with no membrane facing. They should have a Class 25 flame spread and be 1-1/2 inches thick or more.

Installation

General Requirements. Framing and furring systems for use where fire-resistance ratings are required should comply with the governing regulations, and their materials and installation should be identical with assemblies tested and listed by recognized authorities, such as Underwriters Laboratories.

Framing and furring systems for the support of gypsum board products should at least comply with the applicable requirements of ASTM standards C 754 and C 840. In addition, load-bearing framing to receive gypsum board products should be installed in compliance with ASTM Standard C 1007. Gypsum board manufacturers may, however, have their own, more stringent requirements.

Non–load bearing steel studs for standard-thickness plaster (1/2 inch or more) should be installed in accordance with ASTM Standard C 754.

Load-bearing steel studs for standard-thickness plaster should be installed in accordance with ASTM Standard C 1007.

Furring for interior standard-thickness gypsum plaster should be installed in accordance with ASTM Standard C 841.

Furring for portland cement plaster should be installed in accordance with ASTM Standard C 1063.

In addition, light steel framing and furring for plaster to be installed using metal lath should be installed in compliance with the Metal Lath/Steel Framing Association's *Specifications for Metal Lathing and Furring*.

Framing and furring that abuts vertical or overhead building structural elements should be isolated from structural movement sufficiently to prevent the transfer of loads into framing or furring. Open spaces and resilient fillers are often used to fill spaces to prevent load transfer. Care must be taken to ensure that lateral support is maintained.

Both sides of control and expansion joints should be framed independently. Such joints should not be bridged by framing or furring members.

The flat surfaces of framing and furring should not be warped, bowed, or out of plumb or level by more than 1/8 inch in 12 feet in both directions, by more than 1/16 inch across any joint, or by 1/8 inch total along any single member.

On exterior walls a sheet of 15-pound asphalt felt should be placed between each metal framing or furring member and the interior face of the exterior wall.

Metal Furring.   Metal furring should be installed, as appropriate, to support nonstructural metal items. It should be placed on the lines and levels necessary to cause the metal items to fall into the proper location. It should correct unevenness in a supporting structure or solid substrate. It should also span across existing framing that may be at the wrong spacing to support the new metal item properly.

Metal furring may have been installed to support an existing finish or other item. When appropriate, the same existing furring may also be used to support or fasten a new nonstructural metal item. For example, metal wall furring may have been installed where a solid substrate was not appropriate for direct application of a finish, such as gypsum board or plaster.

Where a thicker furring depth is not required to accommodate insulation or mechanical, electrical, or other devices and metal furring is to be used, hat-shaped furring channels (see Fig. 2-1) are usually used to anchor nonstructural metal items, such as manufactured casework. Hat-shaped furring channels are also usually used to support gypsum board products and other materials that are to be fastened in place using screws. To support gypsum board the channels are usually placed vertically at 16 (sometimes 24) inches on center. Hat-shaped furring channels are usually fastened in place using hammer-set or power-activated stud fasteners or concrete stub nails when the substrate is concrete, solid masonry, stone, ceramic tile, or plaster directly applied over solid bases. Toggle bolts are usually used when the supports are metal or hollow masonry. Fasteners should be staggered on alternate wing flanges of the furring and placed not more than 24 inches on center. End splices should be made by overlapping (nesting) the channels by not less than 8 inches and fastening them together with two fasteners per wing.

Where insulation occurs on the inner surface of an exterior wall, Z-shaped furring members are often used to support and anchor nonstructural metal items and board-type finishes. For nonstructural metal items, these members should be placed at the spacing that is appropriate to support the metal item. For finishes they are usually placed vertically and attached at 16 inches on center. Z's are usually fastened to concrete or masonry using hammer-set or power-activated stud fasteners or concrete stub nails spaced at not more than 24 inches on center. End splices should be made by nesting the Z's no less than 8 inches and fastening the sections together with two fasteners.

Where more furring depth is required, studs of the same or a similar type to those used in partition framing may be used.

Corner furring members are needed when hat-shaped furring channels are used to support finishes. Wood furring is used at jambs and ends where Z-shaped furring members are used and it is not practicable to use Z-shaped furring members.

Wall furring for finishes should extend from the floor to the underside of the structure above. For nonstructural metal items the furring need extend only to the top of the item.

Nonbearing Stud Wall and Partition Framing. Nonbearing stud walls are seldom built just to support nonstructural metal items. Most such items that must stand alone are designed to be self-supporting. New metal items may, however, be fastened to, supported by, or stabilized by existing metal studs, including those used for walls, chases, shafts, wall furring, or furring at columns or pipes.

Typical Nonbearing Metal Stud Walls and Partitions. Except where the ceiling is specifically designed to support the framing, metal partition framing should extend to the underside of the structural slab or structural members above. Framing should be aligned and plumbed accurately. Metal floor and ceiling runners should be securely fastened to cast-in-place concrete slabs using stub nails or power-driven fasteners, and to other substrates with suitable fasteners. They should be located near the ends of the runner members and between them, spaced not more than 24 inches on center.

Metal channel studs should be attached to masonry and concrete walls using power-actuated fasteners or stub nails.

Metal studs should be positioned vertically in floor and ceiling runners will all flanges in a given wall or partition pointing in the same direction at the spacing recommended by the applicable industry standards. Studs adjacent to partition intersections and corners should be anchored to floor and ceiling runner flanges with locking metal fasteners or by positive screw engagement through each stud flange and runner flange. Stud splices should be avoided but when necessary should be made by overlapping (nesting) two studs a minimum of 8 inches and attaching the flanges together with two screws in each flange.

Where metal studs are installed in ceiling runners mounted on the bottom of concrete slabs or other structural elements, a 3/8-inch wide space should be left between the top of the stud and the web of the runner so that loads from deflection of the concrete slab will not be transferred into the studs.

Studs should be located no more than 2 inches from abutting partitions, partition corners, and other construction.

Door and other openings should be framed in accordance with the recommendations of the applicable standards. Double studs and runners should be provided where recommended. All elements should be fastened together and to the adjacent floor and ceiling runners using screws. The openings above opening headers should be framed with short studs. Door frames should be spot grouted at jamb anchor clips using joint compound, plaster, or mortar, depending on the finish.

Chase Walls.   Stud partitions that form chases are usually similar to typical partitions. The maximum width of a chase is controlled by the size and height of the studs used. The width should be in accordance with the applicable standards and the published requirements of the stud and finish manufacturers.

The studs on opposite sides of a chase should be in alignment across the chase with their flanges all pointing in the same direction. The studs should be anchored to the floor and ceiling runner flanges with screws or lock fasteners. Overhead structure deflection should be allowed for in the same way as in a typical partition.

Chases should be braced internally. The type of bracing depends somewhat on the configuration of the chase. Where both sides of a chase are unfaced on the chase side, the braces may be either sections of gypsum board or small (often 2-1/2 inch) metal studs placed at about 48 inches on center vertically. The braces should span horizontally between the studs on opposite sides of the chase and be screw fastened to each stud. Studs used as braces should be fastened with two screws in each end.

When chase wall studs are not opposite each other, 2-1/2 inch or larger continuous runner channels should be applied in the chase across the studs on each side of the chase at about 4-foot intervals vertically and be securely anchored with screws to each stud. The two continuous runners should be at the same level. Cross-braces made of 2-1/2 inch or larger runner sections should then be screw attached to each continuous runner, at the same intervals as the stud spacing.

When one side of a chase is solid or has had a layer of gypsum board applied on the chase side of its studs, one runner should be applied within the chase to the face of the exposed studs and a second runner within the chase to the face of the solid board or other substrate. The two runners should be at the same level. Runners applied at exposed studs should be attached to those studs with screws. Runners applied at studs covered with gypsum board should be fastened through the gypsum board to the underlying studs with screws. Runners applied at solid substrates should be attached with appropriate fasteners. Cross-braces made of 2-1/2 inch or larger runner sections should then be screw attached to each continuous runner at the same intervals as the stud spacing.

Partition Reinforcing.   Additional studs, metal blocking, and plates and headers should be installed within partitions to frame openings and to provide for secure, rigid attachment of both recessed and surface-mounted fixtures, handrails, accessories, equipment, and other nonstructural metal items. The kinds of items requiring additional supports include, but are not limited to, bath and toilet accessories, wall-mounted cabinets, shelving, chalkboards and tackboards, fire-extinguisher cabinets, handrails, and other specialties and equipment. The layout of the channel studs should be coordinated with the requirements for such items. Single or multiple studs should be installed as required for their support. At least two studs should be provided at locations where wall-mounted plumbing-fixture supports are attached.

When existing partitions have not been reinforced to receive new nonstructural metal items such as handrail supports and grab bars, additional supports must be provided. Sometimes the existing stud partition or wall must be partly exposed and new reinforcement installed.

Cold-formed Metal Framing.   Cold-formed metal framing is used when studs or other light framing must bear loads in either the transverse or axial directions. It is often the choice when nonstructural metal items must be supported using light metal framing, especially when the metal item is on the exterior side of an exterior wall, forms an exterior wall, or is located above a roof, where wind loading will be experienced. Cold-formed joists are sometimes used to support metal items that must be suspended when the existing structural system is inappropriate for that purpose. Cold-formed metal framing should be installed in accordance with ASTM Standard C 1007.

Stud Assemblies.   Before cold-formed metal studs and runners are erected they are often preassembled in jigs into panels that are then braced to prevent racking. The completely assembled panels are then lifted into place and fastened to the substrates. The panel size is dictated by the structure, but it should be adjusted to suit the material or item that the assembly will support. Where, for example, plaster (stucco) is the finish over an existing cold-formed metal stud system, the prefabricated panels should have been made smaller than normal and been installed to permit movement between the panels. Plaster cracks may indicate either that the existing panels are too large and have formed a support system that is too rigid for the plaster or that expansion and contraction have not been properly accounted for. New metal panels applied over such construction may also suffer from the system's rigidity. It may be necessary in such a case to take additional precautions to prevent failure of the new metal panels.

Floor and ceiling runner tracks for cold-formed metal stud assemblies should be fastened in place using appropriate fasteners. The fasteners should be placed in accordance with the structural design and the system manufacturer's recommendations, but nails and power-driven fasteners should be placed not more than 24 inches on center. Other fasteners should be placed not more than 16 inches on center.

Panels should be erected with studs set plumb (or in the correct plane), in the proper alignment and spacing, and with the panels plumb (or in the correct plane) and in the proper location and alignment. Stud spacing will often be dictated by structural considerations. When the spacing is not thus dictated it should be in accordance with the recommendations of the manufacturer of the material or item that the studs will support.

Many of the requirements given earlier in this chapter for nonbearing studs apply also to cold-formed metal framing. For example, bracing for fixtures, equipment, services, cabinets, handrails, and similar items must be provided. Differences occur in the thicknesses of the metals used in the framing members and in the methods of fastening used within the various systems.

Screws are used for most connections within nonbearing stud systems. In cold-formed metal framing, connections between like members (two sheet-metal components, for example) are usually welded. Studs are usually welded to floor (bottom) and ceiling (top) runners. Connections between unlike members, as sheet-metal and structural steel, are usually bolted, but even they are sometimes welded. Wire ties should not be used.

Framing members around openings are often heavier and more complex in cold-formed metal framing applications than in nonbearing stud assemblies, because cold-formed assemblies frequently carry loads in addition to their own weight.

Horizontal or diagonal bracing, which is needed only in tall, nonbearing partitions, is usually required with all cold-formed assemblies.

Other Cold-formed Framing.   Cold-formed metal joists should be installed plumb and level, within acceptable tolerances for the finish or item to be supported. Ends should bear not less than 1-1/2 inches on solid supports. Joists' ends should be reinforced. Some of the common methods of doing so include imposing clips, hangers, wood blocking, and sheet-steel channels or angles. Joists that support concentrated loads or crossing loads, such as suspended nonstructural metal items and transverse partitions, may require additional bracing at the point of the applied load. Such bracing might consist of a section of sheet-metal channel or angle placed within the joists vertically. Joists must also be braced and secured to prevent their lateral movement.

Sound Insulation and Caulking.   Sound insulation is sometimes used in walls and partitions to which a new nonstructural item is to be fastened. Where they are used, sound-insulation blankets should be installed over the full height and width of the framing or furring. The process of installing a new metal item should not create gaps in that coverage.

Where gypsum board is the finish, sound insulation in partitions is usually installed between the studs and fastened by staples or adhesives to the concealed face of the gypsum board on one side of the partition. In some cases sound insulation can be friction fit, but in others it may be fastened in place with pins, clips, adhesives, or wire.

Insulation should be tightly butted with its joints in solid contact, without voids. It should be fitted neatly to studs and opening framing and be packed tightly around penetrations and between back-to-back penetrations through walls and partitions. Installed nonstructural metal items should do nothing to defeat the effect of required insulation.

Where sound ratings are required, a 1/4- to 1/2-inch space should be left between partition framing and the adjacent construction when possible. This space should be filled with resilient material and sealed with acoustical sealant. At floor and ceiling runners and end studs that must be held tightly to the adjacent construction, a 1/4- to 3/8-inch-diameter bead of acoustical sealant should be installed at both sides of the runner or stud before the finish is applied. The installation of a new nonstructural metal item should not destroy the effectiveness of an existing acoustical sealant.

Shaft Wall Construction. It is occasionally necessary to fasten a new nonstructural metal item to an existing shaft wall. Shaft walls are used as lightweight fire-resistive enclosure walls for stair, elevator, and dumbwaiter shafts; chutes; and service shafts for air, pipes, conduit, and ducts. They are also used as column fireproofing, fire separation between tenant spaces, and corridor walls.

The term "shaft wall" generally implies a proprietary system of components, including gypsum board and metal framing members furnished by the gypsum board manufacturer. Some are of solid construction, but others are cavity walls. Framing and board application should be in accord with the manufacturer's recommendations to provide the fire rating required by code and the desired sound transmission class. Most installations are fire rated for two hours or more.

Shaft wall assemblies and each individual component should be rated by Underwriters Laboratories or another recognized agency at the required fire-resistance and sound-transmission levels. Structural performance must be adequate for the purpose used.

Shaft walls may be finished with painted or otherwise decorated gypsum board, veneer plaster, or ceramic tile over water-resistant gypsum board.

Many of the requirements for other stud walls and partitions apply also to shaft wall systems. For example, isolation from the structure and supplementary blocking and framing to support applied items such as cabinets and toilet fixtures are required.

Metal Framing and Furring Failures and What to Do about Them

Metal Framing and Furring Problems

Metal framing and furring systems may fail because of the types of concrete or steel structure failure or movement discussed earlier in this chapter or the solid substrate problems or other building element problems discussed later in this chapter. Metal furring applied over wood framing may fail because of a failure in the wood structure, as discussed later in this chapter. In addition, framing and furring may fail because of problems inherent in the framing and furring systems themselves.

Metal framing and furring problems that can cause nonstructural metal unit failures include those discussed in the following paragraphs.

Metal framing or furring that is out of alignment or placed too far apart can cause an attached nonstructural metal item to fail.

If the support system itself should fail, whatever the cause, including losing its stability or separating from the substrates, the failure of a supported nonstructural metal item is inevitable.

Metal framing or furring installed in such a way that unplanned-for loads from a deflecting or otherwise moving structure are passed into the framing or furring, and subsequently into the nonstructural metal item, will cause the nonstructural metal item to fail. One way to prevent stress from being applied to wall or partition framing or applied items is to provide a space at the top of the framing, and then fill it with a compressible material. The lateral stability of the wall or partition must, of course, be maintained. Partition and wall furring should be held away from adjacent solid substrates and structural elements.

Welded stud systems that are installed too rigidly, with prefabricated panels that are too large, and with insufficient provision for movement will cause applied nonstructural metal items to fail.

Repairing and Extending Metal Framing and Furring

It may not be feasible to repair failed metal framing or furring members. Even when doing so is possible it is seldom practicable, for example, to clean and repaint severely rusted furring channels. So most of the time when this text refers to repairing framing or furring it means removing the damaged pieces and installing new ones in their place.

The following discussion assumes that structure and solid substrate damage has been repaired and that there is satisfactory support for the framing or furring that is to be repaired.

Where existing framing or furring in a system having a fire or sound rating is to be altered, patched, or extended, and in other locations where assemblies with fire-resistance ratings are required to comply with governing regulations, the materials and installation should be identical with applicable assemblies tested and listed by recognized authorities and in compliance with the requirements of the building code. Materials for use in existing fire-rated assemblies should, unless doing so violates what has just been said, exactly match the materials in the existing fire-rated assembly.

Materials.   The materials used to repair existing metal framing or furring should match those in place as nearly as possible but should not be inferior in quality, size, or type than those recommended by recognized authorities or required by the building code. Errors in the original application should not be duplicated purely for the sake of matching the original.

Damaged existing materials should not be used in making repairs or in extending the existing surfaces.

Repairs. In general, repairs should be made in accordance with recognized standards, such as those mentioned earlier or in "Where to Get More Information" at the end of this chapter, or both, and in the standards referenced in the Bibliography as being applicable to this chapter.

The following discussion contains some generally accepted suggestions for repairing metal framing and furring systems. Because the suggestions apply to many situations, they might not apply in a specific case. In addition, there are many possible cases not specifically covered here. When a condition arises in the field that is not addressed here, advice should be sought from the additional sources of data mentioned in this book. Often, consultation with the manufacturer of the finish or item being supported will help. Sometimes, however, it is necessary to seek professional help (see Chapter 1). Under no circumstances should the specific recommendations in this chapter be followed without careful investigation and the application of professional expertise and judgment.

Preparation. Where existing framing or furring members are damaged, the covering plaster and lath, gypsum board, tile, or other finish must be removed enough to permit the repairs to be made. Where the failure of an existing nonstructural metal item was due to failure of part of a framing or furring system, it may also be necessary to remove the metal item before repairing the failed framing or furring. After the damaged framing or furring has been exposed, it should be repaired in place. If in-place repair is not possible, the damaged framing or furring should be removed, along with its attachments. Damaged elements that are removed should be discarded. These include but are not limited to framing or furring members that are broken, bent, or otherwise physically damaged, rusted beyond repair, or otherwise unsuitable for reuse, as well as attachments that are rusted, broken, or otherwise damaged. Existing attachments that are sound, adequate, and suitable for reuse may either be left in place or removed, cleaned, and reused.

Misaligned, Bent, or Twisted Framing or Furring. Even metal framing of furring that is misaligned, bent, or twisted enough to cause damage to an applied nonstructural metal item can sometimes be left in place and the metal item be reattached so as to overcome the poor condition of the framing or furring. When repair and reapplication of the attached item alone will not prevent failure from recurring, it will be necessary to remove the damaged framing or furring and provide new materials. Sometimes removed metal framing or furring that was only misaligned can be properly reinstalled.

Incorrectly Constructed Framing or Furring. When existing framing or furring was built in such a way that the finish or applied metal item has become damaged, the necessary corrective measures depend on the type of error. The possibilities are many, the solutions even more numerous.

Each case must be examined to determine the true cause, before steps are taken to correct supposed errors. The extent of the damage must also be taken into account. If it is possible to prevent failure from recurring simply by refastening the finish or metal item, for example, this should be done before expensive reconstruction of the framing or furring is undertaken.

Installing New Metal Framing and Furring over Existing Materials

Where a new nonstructural metal item is to be applied over an existing material, a new framing or furring system may be needed. Installing a new structural framing system and repairing and extending an existing structural framing system are both beyond the scope of this book. The principles for installing the types of metal framing and furring systems discussed in this book in existing construction are similar to those for installing new framing and furring systems.

Except for a few different requirements related to fastening both bearing stud walls and nonbearing stud partitions to existing construction, and the possible necessity of removing some existing construction in preparation for their installation, the same requirements apply to both of them in new and existing buildings.

Even in new buildings, furring is always applied over something else. The problems vary only slightly when the "something else" is old. More shimming may be necessary when new furring is applied on existing substrates, of course, and sometimes an existing surface is in such poor shape that it cannot be satisfactorily furred. When this happens it may be necessary to fur the surface with an independent partition or remove the existing construction completely and build it anew.

It is not always necessary to remove existing trim at openings when installing a new furring system over an existing surface. Items that will be concealed in the finished installation can be left in place.

Wood Framing and Furring

Some nonstructural metal items, such as steel stairs, are not usually found as part of the original construction in wood-framed buildings, but other nonstructural items fabricated from metal are routinely fastened to or supported by wood framing and furring in both existing and new buildings.

Wood framing to which nonstructural metal items are fastened or that supports such items includes portions of the structure such as joints, trusses, beams, columns, and bearing studs, as well as partition framing. Though there is no attempt made here to discuss in detail the many possible building framing systems, the general requirements related to stability, tolerances, and materials included here apply to the structural portions of a building when they anchor or support nonstructural metal items.

Most wood structural framing systems in modern nonresidential buildings are supported on either concrete or masonry piers or walls that are in turn supported on concrete footings. Such buildings often have a concrete slab as the first floor. Older buildings are more likely to have a wood floor over a basement or crawl space. Older buildings may also be supported on masonry piers or walls set on concrete footings. Sometimes the footings are of stone or masonry. The number of possibilities is large. Failure in the supporting walls, piers, or foundations can drastically affect the building, including applied or supported nonstructural metal items.

When this text refers to a wood structural system in the broad sense, it means to include the supporting construction and footings that are a part of a wood-framed building.

Wood framing systems may contain some structural steel components. Steel structural elements, such as beams and columns, are frequently used where long spans are desired in wood-framed buildings. The discussion under "Concrete and Steel Structural Framing Systems" earlier in this chapter applies to the steel components of wood-framed buildings.

Wood furring is often used to support and anchor some types of nonstructural metal items such as lockers, for example, even in steel- and concrete-framed buildings.

General Requirements

Wood framing and furring should comply with the applicable building codes and the standards and minimum requirements of generally recognized industry standards, such as the following:

- The American Institute of Timber Construction (AITC), *Timber Construction Standards* and *Timber Construction Manual*.

- The U.S. Department of Commerce (DOC), *PS 1—Construction and Industrial Plywood* and *PS 20—American Softwood Lumber Standard.*
- The National Forest Products Association (NFPA), *National Design Specifications for Wood Construction, Span Tables for Joists and Rafters,* and *Manual for House Framing.*
- The Southern Pine Inspection Bureau (SPIB), *Standard Grading Rules for Southern Pine Lumber.*
- The Western Wood Products Association (WWPA), *Grading Rules for Western Lumber, Grade Stamp Manual, A-2: Lumber Specifications Information, Western Woods Use Book,* and *Wood Frame Design.*
- The West Coast Lumber Inspection Bureau (WCLIB), *Standard Grading Rules for West Coast Lumber.*
- Applicable American Wood-Preservers Association (AWPA) standards.
- Applicable federal specifications and ASTM standards.
- Applicable rules of the respective grading and inspecting agencies for the species and products indicated.

Wood and Plywood

Moisture Content of Lumber.   One cause of problems with nonstructural metals fastened to or supported by wood framing or furring is the use of improperly seasoned (cured) wood. Wood for framing and furring should be seasoned lumber with a 19 percent maximum moisture content at the time of dressing. Lumber with a higher moisture content can be expected to change in size by 1 percent for each 4 percent increase in moisture content. As the wood changes in size it will usually warp and twist, especially when held in place at the ends, as are framing and furring members.

Lumber Classification and Grades.   Softwood materials should comply with the U.S. Department of Commerce's *PS 20* and the National Forest Products Association's *National Design Specifications for Wood Construction.* Each piece of lumber should be grade stamped by an agency certified by the Board of Review of the American Lumber Standards Committee.

Studs, joists, rafters, foundation plates and sills, planking, beams, stringers, posts, structural sheathing, and similar load-bearing members should be of at least the minimum grades required by the applicable building code and the engineering calculations made for the building. For example, light framing lumber might be "Stud" or "Standard" grade lumber for studs and "Standard" grade for other light framing, or "Construction," "Standard," or "Utility" grade for all light framing. Heavier structural framing might be "Select Structural Grade" or "Southern Pine No. 1 Dense" grade.

Often, structural lumber is selected by its structural characteristics. It might, for example, have the following minimum allowable stresses and modulus of elasticity:

Fb (extreme fiber stress in bending): 1,250 pounds per square inch.

Fv (horizontal shear): 95 pounds per square inch.

Fc (compression perpendicular to grain): 385 pounds per square inch.

E (modulus of elasticity): 1,500,000 pounds per square inch.

Blocking, nailers, and similar items are usually the same grade as the framing.

Rough carpentry boards might be WWPA "No. 2 Common," or SPIB "Southern Pine No. 2."

Concealed trim and blocking might be Douglas fir "C Select" or equivalent WWPA softwood.

The previous are examples only; the actual lumber used may have different characteristics.

Lumber Species.   Lumber for framing and furring may be from any of a number of available species, including but not limited to Douglas fir, Douglas-fir-larch, hem-fir, southern pine, spruce-pine-fir, and redwood.

Lumber Sizes.   In newer buildings actual wood sizes will probably be in accord with the U.S. Department of Commerce's *PS 20*, and the lumber will probably be surfaced on all four sides. In older buildings sizes may be actual (a full two inches by four inches instead of 1-1/2 by 3-1/2 inches) or a different actual dimension for the same nominal size. For many years the normal actual size for a 2 by piece of lumber was 1-5/8 inches, not today's 1-1/2 inches, for example. Older furring and framing lumber may not be surfaced on any sides or may be surfaced only where finishes are applied.

Softwood Plywood.   While not usually used in furring systems, softwood plywood might appear in some framing systems. Much softwood plywood used in framing is Douglas fir manufactured and graded in accordance with the U.S. Department of Commerce's *PS 1*. Each panel should bear the appropriate American Plywood Association (APA) grade trademark, indicating the grade, but the grade mark may not be visible after installation.

Miscellaneous Framing Components and Other Materials

Wood Roof Trusses. Trusses should have been designed for the manufacturer by a registered professional engineer to support the loads to be applied and in accordance with building code requirements.

Roof trusses are factory fabricated, often from No. 2 Douglas fir or its equivalent. Connections are usually made with metal plate connectors or plywood gussets. Metal truss plates, whether they are pressed-in tooth or nail-in types, should comply with the Truss Plate Institute's "Design Specifications for Light Metal Plate Connected Wood Trusses." Truss-plate connectors less than 1/8 inch thick should be zinc coated or of noncorrosive metal and conform to the provisions of ASTM Standard A 525 for the Commercial Coating class.

Wood Truss-Type Floor Joists. Joists should have been designed for the manufacturer by a registered engineer to support the loads to be applied and in accordance with applicable building code requirements. These are proprietary products with unique characteristics. Where it is necessary to work with or alter them their manufacturer should be ascertained and consulted in advance.

Miscellaneous Materials. The following miscellaneous materials are necessary for wood framing and furring installation:

Rough hardware, metal fasteners, supports, and anchors.

- Nails, spikes, screws, bolts, clips, anchors, and similar items of sizes and types to secure framing and furring members rigidly in place. These items should be hot-dip galvanized or plated when they will be in contact with concrete, masonry, or pessure-treated wood or plywood and where they will be subject to high moisture conditions or exposed.
- Suitable rough and finish hardware as necessary.
- Bolts, toggle bolts, sheet-metal screws, and other suitable approved anchors and fasteners. These should be located not more than 36 inches on center to firmly secure wood furring, plates, nailers, blocking, grounds, and other wood members and plywood in place. Nuts and washers should be included with each bolt. Anchors and fasteners set in concrete or masonry should be hot-dip galvanized and be of types designed to be embedded in concrete or masonry, as applicable, and to form a permanent anchorage.
- Special fasteners or framing devices are sometimes used in lieu of conventional fasteners such as nails, screws, and bolts to improve installa-

tion procedures or provide higher load values. Such devices should be designed and manufactured to conform with pertinent standards for material selection and performance. Where framing or furring fastened with such devices must be altered or modified, the manufacturer should be consulted for particulars, including bearing and load-carrying capacities and methods to be used in dealing with the devices. Such devices should have the manufacturer's name or identifying mark on them.

■ Building paper is often placed between framing and furring and the substrate, especially at exterior walls, to prevent moisture transfer from the substrate to the framing or furring. The material used most is 15-pound asphalt-saturated unperforated felt complying with ASTM Standard D 226.

Wood and Plywood Treatment

Preservative Treatment. Treated lumber or plywood should comply with the applicable standards of the American Wood Preservers Association (AWPA) and the American Wood Preservers Bureau (AWPB). Each treated item should be marked with an AWPB "Quality Mark," but the markings may not be visible after installation.

Most recently treated framing and furring will have been treated in compliance with AWPA standards C2 for lumber and C9 for plywood, using water-borne preservatives complying with AWPB Standard LP-2. Some installations might be treated using other AWPA- or AWPB-recognized chemicals. Older applications may have been treated in accordance with other standards using other chemicals, sometimes even creosote.

After treatment, treated wood should be kiln dried to a maximum moisture content of 19 percent.

Most framing and furring will not be pressure-preservative treated. Treated portions of framing and furring usually include wood and plywood near or in contact with roofing or associated flashing, and wood sills, sleepers, blocking, furring, stripping, foundation plates, and other concealed members in contact with masonry or concrete.

Treated items should be fabricated before treatment when possible. When treated material is later cut, the cut surfaces should be coated with a heavy brush coat of the same chemical used for the initial treatment.

Fire-retardant Treatment. Fire-retardant treated lumber and plywood should comply with AWPA standards for pressure impregnation with fire-retardant chemicals. They should have a flame-spread rating of not more than 25 when tested in accordance with Underwriters Laboratories Incorporated's (UL) Test 723 or ASTM Standard E 84. The treated material should show no increase in flame spread and no significant progressive combustion when the test is continued for twenty minutes longer than those standards require. The materials used should not have a deleterious effect on connectors or fasteners.

Treated items that are to be exposed to the exterior or to high humidities should be treated with materials that show no change in their fire-hazard classification when subjected to the standard rain test stipulated in UL Test 790.

Fire-retardant treatment chemicals should not bleed through or adversely affect the type of finish used.

Each piece of fire-retardant treated lumber and plywood should have a UL label, but the labels may not be visible after application.

After treatment, treated wood should be kiln dried to a maximum moisture content of 19 percent.

Framing and furring in walls and floors where fire rated construction is required by the building code will probably be treated.

Installation

General Requirements. Lumber and plywood with defects that might impair the quality of the finished surfaces or that are too small to use in fabricating the framing or furring with the minimum number of joints possible or with the optimum joint arrangement should not be used.

The framing or furring should be laid out carefully and set accurately to the correct levels and lines. Members should be plumb, true, and accurately cut and fitted. Openings should be framed and blocking provided for the related work of other trades.

Framing materials should be sorted so that defects will have the least detrimental effect on the stability and appearance of the installation. Large or unsound knots should be avoided at connections, and materials at corners should be straight.

Framing and furring should be securely attached to the substrates in the proper locations and should be level, plumb, square, and in line. Anchoring and fastening should be done in accordance with applicable recognized standards. Nailing, for example, should be done in accordance with the "Recommended Nailing Schedule" in the National Forest Products Association's *Manual for House Framing*. Flat surfaces should not be warped, bowed, or out of plumb, level, or alignment with adjacent pieces by more than 1/8 inch in every eight feet.

Nailing should be done with common wire nails. Fasteners should be of lengths short enough not to penetrate members where the opposite side will be exposed to view or will receive finish materials. Connections between members should be made tight. Fasteners should be installed without splitting the wood, using predrilling if necessary. Work should be braced to hold it in proper position, nails and spikes should be driven home, and bolt nuts should be pulled up tight, with their heads and washers in contact with the work. Shims and wedges should be avoided.

Framing and furring should be spiked and nailed using the largest practicable sizes of spikes and nails. The recommendations of the applicable recognized standards should be followed.

Plywood should be installed in conformance with the recommendations of the American Plywood Association.

Framing and furring that abuts vertical or overhead building structural elements that they are not intended to support should be isolated sufficiently from structural movement to prevent the transfer of loads into the framing or furring. Open spaces and resilient fillers are often used to fill spaces and prevent load transfer. Care must be taken to ensure that lateral support is maintained. Such isolation is particularly a problem when the main structural system is steel or concrete.

Both sides of control and expansion joints should be framed independently. The joints should not be bridged by framing or furring members.

Anchors and Fastening Systems.   Bolts, lag screws, and other anchors should be used to anchor framing and furring in place. Generally, fasteners are placed near the tops and bottoms or ends of items and not more than 36 inches on center between them. Shorter members should, however, be anchored at 30 inches on center.

Bolts should have nuts and washers.

Anchor bolts should be not less than 1/2 inch in diameter, with the wall end bent 2 inches. They should extend not less than 8 inches into concrete and 15 inches into grouted masonry units. They should be placed at 48 inches on center, with not fewer than two bolts in each member.

Expansion bolts should be not less than 1/2 inch in diameter and be placed into expansion shields. The expansion shields should be accurately recessed at least 2-1/2 inches into concrete.

Exterior wall sills should be anchored with anchor bolts as indicated above.

Interior bearing wall sills should be anchored, using expansion bolts, into concrete at 48 inches on center, with not fewer than two bolts in each member; or with shot pins with cadmium washers into concrete 6 inches from corners and splices and not less than 36 inches on center, using no fewer than two pins in each member.

Interior nonbearing partition sills should be installed as described in the previous paragraph for bearing walls, except that the shot pins' spacing may be increased to 48 inches on center.

Bolts in wood framing should be standard machine bolts with standard malleable iron washers or steel-plate washers. Steel-plate washer sizes should be about 2-1/2 inches square by 5/16 inch thick for 1/2- and 5/8-inch diameter bolts, or 2-5/8 inches square by 5/16 inch thick for 3/4-inch diameter bolts. Bolt holes in wood should be drilled 1/16 inch larger than the bolt's diameter.

Lag bolts should be square headed and made of structural-grade steel. Washers should be placed under the heads of lag bolts bearing on wood.

Framing and Furring Spacings. The spacing of framing or furring members installed specifically to fasten or support nonstructural metal items will be dictated, of course, by the requirements of the item being fastened or supported, but spacing may vary depending also on the underlying construction. The spacing of underlying framing or furring must be considered when a new nonstructural metal item to be installed in an existing space must be fastened to, or be supported by, the existing construction.

Spacing requirements are generally the same for both framing or furring, but they may vary in practice with the substrate and the material being supported. Structural considerations may dictate the spacing of framing. For example, load-bearing stud spacings will be determined based on the loads to be applied and the studs' sizes and spans. Sometimes even non-load-bearing spacings will be determined by stud lengths, but more often the spacings will be maintained and the studs increased in size to accommodate increased stud lengths.

The spacing of framing and furring that has been installed to support finishes or existing items of other types may vary considerably from building to building, or even within the same building. Depending on the conditions and the finish to be applied, spacings may range from as narrow as 12 to as wide as 60 inches. In general, framing or furring that is supporting gypsum board products on walls will be either 16 or 24 inches on center. That supporting metal plaster lath on walls will be either 12, 16, 19, or 24 inches on center. That supporing ceramic tile will vary depending on the substrate material, but the spacing should not exceed 16 inches on center. Other spacings may of course be found in an existing building.

The book in this series entitled *Repairing and Extending Finishes: Part I* contains an extensive description of the spacings that should be found in framing or furring when the finish is plaster, gypsum board, or ceramic tile. Many of the documents listed in the "Where to Get More Information" section at the end of this chapter contain recommendations related to framing and furring spacings to support the particular finish material they cover.

The manufacturers of the finishes and framing and furring are also sources of advice about spacings that might have been used in an existing installation.

When an installation has spacings that vary from those recommended by the industry this does not automatically mean that failure will occur. When failure has occurred, however, deviation from industry recommendations should be examined as a potential cause.

Wood Furring.   Appropriate wood furring should be installed to support nonstructural metal items. It should be placed on the lines and levels necessary to cause the metal items to fall into the proper locations. It should also correct unevenness in a supporting structure or substrate.

The size of furring lumber should be that recommended by the manufacturer or fabricator of the metal item that will be attached, but it should not be less than 1-1/2 inches wide (2-inch nominal) lumber. Ordinarily, lumber that is 1 inch in nominal thickness (3/4-inch actual) may be used over a solid substrate, but not over framing. Furring over framing should be at least 1-1/2 inches thick (2-inch nominal), to prevent flexing of the members. Of course, furring members in existing construction may be of almost any size that was available at the time.

Furring should form a complete system adequate to properly support the metal item that will be attached.

Wood Framing.   Framing includes wood stud walls and partitions; floor and roof framing; columns, posts, beams, and girders; and trusses. In general, framing design and member sizes, spacings, and locations should comply with recognized standards such as the National Forest Products Association's *Manual for House Framing.*

Framing should be anchored, tied, and braced in such a way that it will develop the strength and rigidity necessary for its purpose. Members should not be spliced between supports.

The information that follows is not intended to give enough information to build wood framing. The intention is rather to highlight those aspects of wood framing that, if not properly done, might cause a supported or attached nonstructural metal item to fail. Refer to the sources in "Where to Get More Information" at the end of this chapter for detailed data about constructing wood framing.

Stud Walls and Partitions.   Sills should be fastened securely in place. Where subflooring occurs, the sills should be placed over the subflooring.

Usually, interior stud partitions are framed with 2 by 4 lumber, but sometimes 2 by 3's are used. Exterior walls and some interior bearing walls are often framed with 2 by 6 lumber. Partitions containing pipes are also sometimes framed with 2 by 6 lumber.

Interior wall and partition studs are ordinarily placed 16 inches on center. Exterior wall studs are usually placed either at 16 or 24 inches on center, depending on the wall's construction and the stud sizes.

Studs should be nailed securely to a sill plate and a top plate. In bearing walls, after the wall framing has been secured in place a second top plate should be added and nailed securely to the first top plate. Double top plates are also used sometimes in nonbearing walls and partitions.

Walls and partitions should be set level, plumb, true to line, in the proper location, and be properly braced to prevent later movement. Members in walls and partitions should also be in alignment with each other and be free from twist or warp. Their faces should be flush with each other.

Every opening should be framed with at least two studs at the jambs and a wood lintel consisting of as many members as necessary to ensure that the lintel finishes flush with the studs on each side of the wall. Three or more studs may be needed at the jambs of wide openings, especially in load-bearing walls.

The bottom of each window opening should have a rough sill consisting of two stud-sized horizontal members.

Where plumbing, heating, or other pipes occur in walls and partitions, studs should be placed to accommodate the pipes while still giving proper support for the finishes.

Bridging should be provided in stud walls and partitions where recommended, to produce a stable wall or partition.

Corners and intersections should be framed with at least three studs, to produce a stable corner or intersection and provide appropriate fastening surfaces for the finish material.

Loads imposed by structural elements as they deflect should be prevented from being passed down into stud walls and partitions, except when a wood-framed wall is supposed to carry the load from above. One way to prevent stress from being applied to the wall or partition framing is to provide a space at the top of the framing, which is then filled with a compressible material. The lateral stability of the wall or partition must, of course, be maintained.

Other Structural Framing Elements.   Nonstructural metal items are often fastened directly to wood floor and ceiling joists, roof trusses, columns, beams, and girders. To provide an appropriate base for fastening metal items such framing should be level or in plane, stable, plumb, true, securely fastened in place, and properly spaced.

Where pipes or other items interrupt the spacing of framing components, additional framing or furring must be provided to stabilize the framing and support the applied metal items properly.

Framing must be doubled or otherwise increased beneath unusual loads, to provide a properly stable and level surface for the metal items to be applied.

Bridging should be provided in floor and roof framing to ensure stability and help prevent warp or twist in framing members.

Wood Framing and Furring Failures and What to Do about Them

Failures in a wood framing system or wood furring can affect applied or supported nonstructural metal items in two ways: the structure or furring can fail, or the structure or furring can move in ways not anticipated in the nonstructural metal item's design.

Wood Framing and Furring Problems

Wood framing and furring that is supported by a concrete or steel structural framing system may fail because of structure failure or movement, as discussed earlier in this chapter. Wood framing or furring supported by a solid substrate may fail because of solid substrate problems, as discussed later in this chapter. Such wood framing or furring may also fail because of problems inherent in the framing and furring itself.

Wood framing that is a building's structural system may fail because of problems with the foundations or other supports, or because of problems with the framing itself.

Problems that can cause failure in a nonstructural metal item attached to or supported by wood framing or furring include those addressed in the following paragraphs.

Wood framing or furring may fail due to failure of the concrete or steel structure that supports the wood framing or furring. Refer to "Concrete and Steel Structure Failure" earlier in this chapter for a full discussion. The reasons for failure listed there apply also to the concrete and steel portions of a wood-framed building.

Excess or unprepared-for movement in a concrete or steel structure that supports the wood framing or furring can cause the wood framing or furring to fail. Refer to "Concrete and Steel Structure Movement" earlier in this chapter for a complete discussion. The reasons for failure listed there apply equally to the concrete and steel portions of a wood-framed building.

Problems with a solid substrate that is supporting wood furring can cause the furring to fail. Refer to "Solid Substrate Problems" later in this chapter.

Problems with other building elements can also cause wood framing or furring to fail. Refer to "Other Building Element Problems" later in this chapter for a full discussion.

Wood framing or furring may fail because it has been improperly designed or because it experiences unanticipated conditions that exceed the design limitations. Design limitations are dictated by the materials' characteristics, legal requirements, and economic factors. It is not economically feasible to design every structure to handle every condition that might occur.

Wood structure failure may be minor or extensive in scope, from slight damage to complete building collapse, but most wood structure failure is relatively small in magnitude. A single cause may generate failure at any level. For example, an undersized footing may lead to building collapse— or simply to more settlement than normal. Wood structural-system failure will probably be a result of one of the following causes:

- The structure was not properly designed to withstand all loads to be applied, without excess deflection, vibration, settlement (especially differential settlement), expansion, or contraction.
- The building's foundations were not properly designed or installed, resulting in excessive or uneven settlement.
- An unforeseen traumatic event occurred that applied forces to the building that it was not designed to resist. Earthquakes, hurricanes, and tornadoes (Fig. 2-5) are examples.

Damage to nonstructural metals items can occur if the structure moves, especially if the movement is larger than expected. Structure movement should be suspected when a nonstructural metal item fails, especially if the evidence of failure is broken or open joints or an object fabricated from nonstructural metal separating from adjoining construction. To prevent failure, the design, fabrication, and installation of nonstructural metal items must take the possibility of structure movement into account.

Undue structure movement may be a symptom of structural failure. Some structure movement is, however, normal and unavoidable. Expected structural movement may be caused by wind, thermal expansion and contraction, and deflection under loads. Normal movement in a wood structure may be caused by one of the following:

- Variable wind pressure.
- Structure settlement.

Figure 2-5   A tornado collapsed this small wood-framed building. It is not possible to prevent this kind of damage, which will, of course, usually destroy all associated nonstructural metal items.

- Thermal expansion and contraction.
- Deflection of structural members. Deflection is usually a temporary condition. When the loads are removed, structural elements often return to their original shape. Permanent deflection can occur, however, if the applied loads are large.
- Structure vibration, which is often transferred from operating equipment in the building. Vibration can loosen fasteners.

Misaligned, twisted, or protruding wood framing or furring can damage a supported or attached nonstructural metal item. Where these occur the metal item may be forced out of alignment or position. Such defects may be caused by changes in lumber's size. Even relatively dry wood will shrink, which tends to cause the lumber to warp or twist. Lumber may also expand, from absorbing free water, condensation, or water vapor in high-humidity conditions.

Failure may occur when supports that are placed too far apart do not offer sufficient support for an attached nonstructural metal item.

When a support system loses stability or separates from its substrates, the failure of an attached or supported nonstructural metal item may follow.

Wood framing or furring may be installed in such a way that loads from a deflecting or otherwise moving structure are passed into the framing or furring and subsequently into a supported nonstructural metal item, causing it to fail. Examples of errors that can result in such loads being transferred into the framing or furring include wedging the ends of furring against masonry or concrete, building the ends of furring members into masonry, and building a stud partition tightly against the bottom of a concrete slab.

Repairing and Extending Wood Framing and Furring

Even minor failure in a wood structure may damage supported nonstructural metal items; thus, before repairing failed nonstructural metal items it is necessary to determine whether structural failure is responsible for the metal item's failing. Where structural failure is to blame, it is usually necessary to correct the structural failure before repairing the nonstructural metal item. Otherwise the failure will usually recur. In severe cases, structural reconstruction such as shoring up beams, adding columns, or replacing structural members may be necessary. When, however, the structure damage is self-limiting and not dangerous to the building or to people it is sometimes possible to modify the existing support system or provide a new one for the nonstructural metal without making major corrections to the failed structure.

Just as the construction of new wood structural systems is not covered in detail in this book, detailed recommendations for major repairs to them are not included here. Although minor repairs to wood structural elements are touched on and wood studs are discussed at some length, this book has in general been written with the assumption that major failures in wood structural elements, if any, have been diagnosed and necessary repairs have been made.

Most failed wood framing or furring members are removed and new material installed, because even when it is possible to repair damaged wood framing or furring members it is not often reasonable to do so. Straightening a warped or twisted stud, for example, will probably prove to be impracticable. Thus, most of the references in the following paragraphs to repairing framing or furring mean removing the damaged pieces and installing new ones in their place.

The following discussion assumes that damaged concrete or steel structural elements and solid substrates supporting the wood framing or furring have been repaired and present a satisfactory support system for the framing or furring being repaired.

Where existing framing or furring in a system having a fire or sound rating is to be altered, patched, or extended, and in other locations where assemblies with fire-resistance ratings are required to comply with governing regulations, the materials and their installation should be identical with applicable assemblies tested and listed by recognized authorities and be in compliance with building-code requirements. Materials for use in existing fire-rated assemblies should exactly match the materials in the existing fire-rated assembly, unless doing so violates this guideline.

Materials. The materials used to repair existing wood framing or furring should match those in place as nearly as possible, but they should not be inferior in quality, size, or type to those recommended by recognized authorities or required by the building code. Where existing materials are fire-retardant treated, new materials must be similarly treated. Where the existing material is pressure-preservative treated, new items should be also.

It is usually best to match lumber sizes exactly when installing a new member in an existing framing or furring system. In older installations, however, the existing lumber may be of sizes that are no longer standard. For example, members that are nominally 2 by 4 may actually be 2 by 4 inches if the building is old, 1-5/8 by 3-5/8 inches as was once the standard, or 1-1/2 by 3-1/2 inches, as is the current standard.

When lumber of the exact size needed to match existing lumber is not available, there are three ways to solve the problem. A larger member can be cut down to the size of existing members. This may be an expensive solution, however. It will usually be less expensive to shim the new member so that its face surfaces are in alignment with the faces of the existing members. The third alternative is to build up a new member from two standard-sized members.

Repairs. In general, repairs should be made in accordance with the recommendations of recognized standards, such as those mentioned earlier in this chapter, or in "Where to Get More Information" at the end of this chapter, or both, and in the standards referenced in the Bibliography as being applicable to this chapter.

The following discussion contains some generally accepted suggestions for repairing wood framing and furring systems. Because the suggestions are meant to apply to many situations, they might not apply in a specific case. In addition, there are many possible situations that are not specifically covered here. When a condition arises in the field that is not addressed in this book, the additional data sources mentioned in this book should be sought out for advice. Often, consultation with the manufacturer of the finish being supported will help. Sometimes it is necessary to obtain professional help (see Chapter 1). Under no circumstances should the specific recommendations in this chapter be followed without careful investigation and the application of professional expertise and judgment.

Preparation. Where existing framing or furring members are damaged, the covering finish (if any) must be removed to the extent necessary to permit repairs to be made. When the framing or furring supports or provides attachment for a nonstructural metal item, it may be necessary to remove the metal item to allow access to the failed framing or furring. When removing the metal item is impracticable, as would be the case with steel stairs, for example, it may be necessary to find another means of access. Sometimes it may be less expensive to construct another support system entirely than to remove the metal item and reinstall it later.

After all obstacles to doing so have been dealt with, damaged existing framing or furring can be removed. Damaged elements that should be removed include wood members that are twisted, warped, broken, rotted, wet, out of alignment, or otherwise unsuitable for use.

Misaligned, Warped, or Twisted Framing or Furring. Even when wood framing has become so misaligned, twisted, or warped that it has damaged an attached or supported nonstructural metal item, it is usually not removed and a new member installed unless the structure itself is in danger. Often such members are simply left in place and the metal item removed and reinstalled in such a manner that the poor condition of the framing is overcome.

It may also be possible to remove, repair, and reinstall a metal item that has failed because furring that supports it or to which it is attached has become misaligned, twisted, or warped. When repair, removal, and reinstallation of the metal item alone will not prevent failure from recurring, however, it will be necessary to remove the damaged furring and install a new piece in its place.

Shrinkage. Where an attached or supported nonstructural metal item becomes misaligned or otherwise fails, due to shrinkage in the lumber used for framing or furring, it is often possible to repair and reattach the metal item without removing the lumber. Where such repairs cannot be accomplished satisfactorily it may be possible to remove all or part of the metal, then shim the framing or furring to produce a flush surface for reinstallation of the metal. Where shimming does not solve the problem, it may be necessary to remove the damaged lumber and provide new framing or furring.

Incorrectly Constructed Framing or Furring. Where the framing or furring was originally built in such a way that a supported or attached nonstructural metal item has become damaged, the necessary corrective measures will depend on the type of error. The possibilities are many, the solutions even more numerous. They range from simply planing down a projecting brace to reconstructing a stud wall or framing. The structural system will probably not need to be remade, however, unless the original error or its consequences has resulted in a dangerous, or at least potentially dangerous, condition.

Each case must be examined to determine the true cause before steps are taken to correct supposed errors. The extent of the damage must also be taken into account. When it is possible to prevent failure from recurring by simply shimming or refastening the nonstructural metal item, for example, this should be done before expensive reconstruction of the framing or furring is undertaken.

Installing New Wood Framing and Furring over Existing Materials

When a new nonstructural metal item is to be applied over an existing material, a new framing or furring system may be needed. Installing new structural framing systems and repairing and extending existing ones are beyond the scope of this book. The principles for installing, in existing construction, new framing and furring systems that are not part of the main structural system are similar to those for installing framing and furring systems in new construction.

Except for a few different requirements related to fastening stud walls and partitions to the existing construction, the same requirements apply to them in new and existing buildings.

Even in new buildings, furring is always applied over something else. The problems vary only slightly when the something else is old. More shimming may be necessary when new furring is applied on existing substrates, of course, and sometimes an existing surface will be in such poor shape that it cannot be satisfactorily furred. When this happens it may be necessary to fur the surface with an independent partition or remove the existing construction completely and build it anew.

Substrates

The substrates beneath nonstructural metal items may be either solid or supported. Supported substrates include plaster, gypsum board, ceramic tile, wood paneling, and other finish materials. Nonstructural metal items are seldom attached to or supported directly by substrates that themselves require a support system. When nonstructural metal items are applied against supported substrates, they should be fastened through the substrates to the framing, furring, or solid substrate underlying the supported substrate. Sometimes a second layer of furring is applied over the supported substrates to support the metal item. The second layer of furring should also be attached to the underlying support system, through the supported substrate material. Framing and furring are addressed in the parts of this chapter entitled "Metal Framing and Furring" and "Wood Framing and Furring."

Solid substrates include those that are part of the structural system and those that are fillers, such as nonbearing walls and partitions. The most common solid substrates to which nonstructural metal items are attached are concrete, concrete masonry, brick, and some types of stone. Metal items may also be fastened to and supported by other types of solid substrates, but more often a separate support system is used for other solid substrates. Some solid substrates should not be used to support loads and therefore should not become supports for load-bearing metal items. Gypsum block falls into the latter category. Solid substrates are usually furred to receive some types of metal items, such as lockers and manufactured casework.

Failures in solid substrates to which metal items are applied, either directly or over furring, will often (though not always) damage the metal item. Some nonstructural items made from metal are, however, virtually independent of their substrates. For example, unless they are quite severe, cracks in a concrete wall behind a row of lockers will probably not damage the lockers. The possibility that solid substrate failure has caused a failure in a nonstructural metal item should be investigated and ruled out, however, before repairs to the metal item are attempted. When a damaged solid substrate is responsible for the failure of a metal item, it is often necessary to repair the solid substrate before repairing the metal item. When the solid substrate's damage is self-limiting and not dangerous to the building or to people, it is sometimes possible to repair an existing furring system or install a new one to support the nonstructural metal item without having to repair the solid substrate.

The repair of solid substrates is beyond the scope of this book, which assumes that if damaged solid substrates have been discovered the necessary repairs have been made. It is not the purpose of this book to discuss solid substrate materials or construction methods. It should be recognized, however, that the design and construction of solid substrates can affect nonstructural metal items that are applied to or supported by them, and their design and construction can contribute to failure of the applied or supported items when it occurs.

Solid Substrate Problems

Solid substrate problems that can lead to nonstructural metal items' failure include those discussed in the following paragraphs.

The solid substrates may exude materials that affect metal or cause its attachments to fail. Some substances extruded by solid substrates that can cause harm to metals are not foreign to the substrate material. It is perfectly natural for a concrete wall to evaporate water for an extended period, for example. Applying a metal item using fasteners that can rust can lead to failure of those fasteners. Wet alkaline materials such as those found in wet mortar will etch aluminum. Problems of this sort are not, however, actually substrate failures; this book calls these kinds of failures bad workmanship.

Some foreign substances that may cause failure in metals, especially those that may affect the finishes used on the metals, however, are the result of bad workmanship in installing the substrate and have nothing to do with the metal item installer's workmanship. Efflorescence is an example.

Attached nonstructural metal items may be damaged if the solid substrate material cracks or breaks up, joints crack, or surfaces spall because of bad materials, incorrect material selection for the location and application, or bad workmanship.

Damage may also occur if the solid substrate material is weaker than required by the standard in accordance with which it was supposed to have been manufactured. This condition can be caused by poor manufacturing techniques or controls. Weak concrete masonry units may not hold fasteners, for example.

Similar problems will occur if a nonstructural metal item is supported from a material that is too weak to support the metal item. Materials that are inherently weak should not be used to support load-bearing metal items. Door head lintels should not be supported on clay tiles or gypsum blocks, for example. This is really a design or workmanship error by those associated with the metal item being supported or attached, however, not the fault of the substrate.

An attached metal item may be damaged if the solid substrate moves excessively from improper design or installation. Types of excess movement that can cause problems include deflection, vibration, settlement (especially differential settlement), expansion, and contraction.

Similar damage may occur if movements normal in a particular solid substrate are not accounted for in the design and installation of a nonstructural metal item. As is true for concrete and steel structural systems, some movement in solid substrates is normal and unavoidable. This movement must be accounted for in the design, fabrication or manufacture, and installation of nonstructural metal items. Normal movement includes that caused by settlement, thermal expansion and contraction, creep, and vibration.

Other Building Elements

Other building elements that are poorly designed or that fail can cause nonstructural metals to fail. These other elements include site grading adjacent to a building; roofing; flashing; waterproofing; insulation; elements that close openings, such as windows, doors, and louvers; caulking and sealants; and mechanical and electrical systems.

Other Building Element Problems

The types of failures in other building elements that can cause nonstructural metal failure include those that follow.

Other building elements, whether they were poorly designed or are simply ones that have failed, that permit water to reach the portions of nonstructural metal items not designed to resist or shed water will damage the metal item or its finish. Such possible sources of water intrusion include roof and plumbing leaks; failed sealants; leaks through doors, windows, louvers, and other devices to seal openings; and leaks through rain gutters and downspouts, including missing ones.

Similar problems will occur if other building elements are designed so that condensation is permitted to form on nonstructural metal items, especially on concealed or unprotected surfaces. Condensation can result from selecting the wrong materials or installation methods for insulation and vapor retarders, or locating those elements improperly. Failing to provide proper ventilation in attic spaces can also lead to condensation. Refer to the book in this series *Repairing and Extending Weather Barriers* for a detailed discussion of condensation and how to prevent it.

Where to Get More Information

The National Forest Products Association's *Manual for House Framing* contains a comprehensive nailing schedule and other significant data about wood framing. It is a useful tool for anyone who must deal with wood construction.

Ramsey/Sleeper's *Architectural Graphic Standards* contains data about nailing arrangements and nail sizes for many framing situations.

The Commerce Publishing Corporation's *The Woodbook* is a wood products reference book published annually. It contains specifications, application recommendations, span tables, and other data about a number of wood products. Most of it consists of fliers and product data published by wood product producers and their associations. Unfortunately, purchasers of *The Woodbook* are expected to pay for this manufacturer's literature, much of which is obtainable free from other sources.

The Forest Products Laboratory's *Handbook No. 72—Wood Handbook* contains a detailed discussion of wood shrinkage.

Some of AIA Service Corporation's *Masterspec* "Basic" sections contain excellent descriptions of the materials and installations that are addressed to the present chapter. Unfortunately, those sections contain little that will help with troubleshooting these installations. Sections that have applicable data are marked with a [2] in the Bibliography.

Of course, every designer should have the full complement of applicable ASTM Standards available for reference, but anyone who needs to understand framing and furring that will support nonstructural metal items, or to which such items will be attached, should have access to a copy of the ASTM Standards marked with a [2] in the Bibliography.

The Gypsum Association has published several documents that include requirements for framing and furring systems to support gypsum board products. Those applicable to the framing and furring discussed in this chapter are marked with a [2] in the Bibliography.

The Metal Lath/Steel Framing Association's publications *Lightweight Steel Framing Systems Manual* and *Specifications for Metal Lathing and Furring* should be available to everyone responsible for metal framing or furring systems.

The United States Gypsum Company's publications *Gypsum Construction Handbook* and *Red Book: Lathing and Plastering Handbook* are excellent data sources for information about the framing and furring that should underlie gypsum board or plaster finishes.

The Western Lath, Plaster, and Drywall Contractors Association (formerly the California Lathing and Plastering Contractor's Association) is responsible for an excellent 1981 publication called *Plaster/Metal Framing System/Lath Manual*. A new edition published in late 1988 by McGraw-Hill was not available at the time of this writing. While the 1981 edition of the book does not cover repairing existing materials, it is a good source of information about metal framing.

Refer also to other items marked [2] in the Bibliography.

Metal Materials

Nonstructural metals used in building and related construction are divided into two broad categories: ferrous metals, meaning all metals that have iron as their basic ingredient, and nonferrous metals, which includes all other metals used.

This chapter discusses materials in both categories. It includes a description of the most-used metal materials, the types of failures that might occur in them, and the industry's recommended methods for dealing with the potential failures.

Chapter 4 discusses the finishes that the metals covered by this chapter have after they leave the milling or casting process, and the mechanical, chemical, and inorganic applied finishes normally used to protect and decorate them.

Chapter 5 describes paint and other organic finishes or metals used in nonstructural metal items.

Chapter 6 covers the design and fabrication of nonstructural metal items. Then Chapter 7 discusses their installation, cleaning, and repair.

For terms used in this chapter, see the meanings listed in the Glossary.

Metals that will be exposed to view in buildings and related construction should be capable of having items produced from them in which the exposed surfaces are flat, smooth, and free from surface blemishes. Their surfaces, when finished, should also not exhibit pitting, seam marks, roller marks, undesirable roughness, "oil canning," stains, discolorations, or other imperfections. After these metals have been formed into nonstructural metal items they should be capable of withstanding the degree of stress of every kind that might be imparted to them, without becoming deformed or damaged in any way.

Iron

Iron is made by removing the oxygen from iron ore in a process called reduction. Several types of ore are used, but they are all forms of ferric oxide. Today this process takes place in a blast furnace. The process is more complicated than this explanation makes it sound, of course. It demands the use of the right fuel (high-quality metallurgical coke), the correct flux (some form of calcium or magnesium carbonate), and the proper furnace design, temperature, and operation.

The reduction of iron ore produces both liquid iron and liquid slag, which is the residue left by the materials used in the process. The liquid iron is either taken directly from the furnace to be used in manufacturing steel or cast into bars called pigs. Liquid iron is the beginning point for all iron-based materials.

The slag is removed periodically from the furnace and permitted to cool. Slag is used for various purposes. For example, it is crushed and used as a roofing aggregate and in concrete. When wrought iron was being made regularly, slag was used in producing it.

Pure iron is a heavy, ductile, malleable metal. When uncorroded it is silver-white in color. In the building industry it is almost never used alone. Pig iron, which looks similar to pure iron, is actually an eutectic alloy containing pure iron and between 2.5 and 4.5 percent carbon. Usually it also contains impurities such as phosphorus, sulfur, silicon, or manganese, which give the iron differing qualities, depending on the type and amount of each impurity present. The composition of pig iron is determined by the type of ore used and the other factors mentioned in the previous paragraph. The primary types of composition are called basic, foundry, malleable, Bessemer, and low phosphorus. Each type is used for a different purpose.

Iron was manufactured far back in the Bronze Age, though not in great amounts. There is evidence, for example, that iron was made by the ancient Egyptians at least as long ago as 4000 B.C. and probably even earlier. Iron was also made in other parts of the world very early in human history. Iron artifacts found in China, for example, date back to 2700 B.C.

By the time of Rameses II, who reigned from 1292 to 1225 B.C., iron was being widely used in Egypt. Over the following thousand years its use spread throughout most of the known world. Iron was not used extensively in western Europe, however, until around 100 B.C.

Wrought Iron

Most early iron objects were made from wrought iron. Until the production of steel by the Bessemer and open-hearth processes became predominant in the 1880s, wrought iron was the primary structural metal used in buildings. Although its popularity diminished, its use for structural purposes continued into this century. New wrought iron is difficult if not impossible to obtain today, but much existing wrought iron will still be found, especially in older buildings and related construction (Fig. 3-1). The material used today in most situations where wrought iron would have been used in the past is a type of low- and intermediate-tensile-strength steel called mild steel (Fig. 3-2). Traditionally, mild steel had a carbon content of between

Figure 3-1   A real wrought iron fence.

Figure 3-2 A mild-steel window grating trying to look like wrought iron.

0.15 and 0.25 percent, but the materials used today to do the jobs once done by wrought iron may vary from that range.

Wrought iron is a two-component material made of almost pure iron, usually containing less than 0.12 percent carbon, and iron silicate (slag). It was originally made directly from ore in a hearth filled with charcoal. Until relatively recently, much of it was made in a process known as puddling. In this process pig iron was placed on a hearth and melted under external heat to the correct consistency, forming a ball of iron with a high slag content, which was then squeezed and rolled. A more modern method for producing wrought iron is the Aston process, which consists of pouring molten low-carbon steel into molten slag at the correct temperature so that the whole mass forms into a granular, spongy ball, which is then squeezed in a hydraulic press to remove excess slag and rolled.

Wrought iron is malleable and ductile at all temperatures. It is relatively soft and easy to forge, machine, and thread. It can also be welded easily, which helps in creating complex shapes. It is resistant to repeated stress and corrosion.

Wrought iron is often thought of today as a metal used primarily in fences, gates, and decorative features (Fig. 3-3), but it may also be found as structural shapes, plates, sheets, tubes, pipes, columns, bars, rods, and fittings in older buildings. It has also been used for bolts and fasteners and in tanks to store various liquids.

Cast Iron

The material called cast iron, or sometimes just iron, is an alloy of iron containing between 2.5 and 4 percent carbon and between 2.25 and 3.75 percent silicon. Cast iron with a carbon content of less than 4.3 percent is called hypoeutectic iron. That with a carbon content in excess of 4.3 percent is called hypereutectic iron (see the definition of "alloy" in the Glossary). Sometimes other metals, such as copper, titanium, chromium, or nickel, are added to create desired chracteristics.

Figure 3-3 Decorative wrought-iron grillage.

The carbon in cast iron may be in the form of either carbide or graphite. Which of these is formed is controlled by the amount of silicon present in the pig iron and its rate of cooling. Cast iron containing carbide is white in appearance and is therefore called white cast iron. Cast iron that contains graphite is gray in color and is consequently called gray iron. White cast iron is extremely hard and brittle. Gray iron is also brittle, but it is stronger than white cast iron. Neither is malleable. White cast iron is usually reserved for use in making malleable iron, described in the following section. Gray iron castings, however, which are usually required to comply with the requirements of ASTM Standard A 48, are far more frequently used in building and related construction than white iron castings.

Cast-iron objects can be made by pouring the molten metal (pig iron) from a blast furnace directly into molds when the pig iron is of the right composition. Normally, however, solid pig iron of the foundry type is remelted to make cast iron. Often scrap iron, scrap steel, or both are added to the foundry pig iron, and the conglomerate is then melted together. The resulting molten cast iron is poured into molds.

Cast iron is used to make a variety of products for use in buildings.

Malleable Iron

The proper name for malleable iron, also called malleable cast iron, is American black heart malleable cast iron. To make it, malleable pig iron and scrap iron are melted together to make white cast iron, which is then poured into a sand mold. After the casting has hardened it is annealed. The resulting material is tough and strong, yet ductile. It is easily machined and can be cast into complicated shapes. A slight change in the production process results in a material known as pearlitic malleable iron, which is much stronger than ordinary malleable iron.

Malleable iron is used as fitting for pipes, for manufacturing various types of machinery, and for manufacturing builders' hardware. Malleable-iron castings should comply with ASTM Standard A 47.

Other Types of Cast Iron

There are other types of cast iron that are occasionally used to make products used in the building industry. They include alloy cast iron, which contains such materials as chromium, nickel, and molybdenum to give it special characteristics.

Another special cast-iron material is nodular cast iron, which is very strong, ductile, corrosion resistant, machinable, and tough. It is produced by adding magnesium to cast iron.

Iron Castings

Iron castings are generally cheaper than similarly shaped rolled, milled, or forged steel components. They are also cheaper than steel castings, because they require lower temperatures to produce. Gray-iron castings are cheap, strong, and easily machined.

Traditional static molds for iron casting are made by machine, using foundry sand. Modern molds may also be produced from sands containing cement or other materials. Cores are placed in the molds and molten iron poured into the space between the sand and the core. The cores, which are burned away by the hot metal, collapse when the metal cools. A process called "chill casting" is also sometimes used in which the mold is cooled, causing the carbon in the gray iron to convert chemically to iron carbide (see the section on "Cast Iron" earlier in this chapter). The result is a gray-iron core surrounded by white iron. Thus, the molding has both the strength of gray cast iron and the surface hardness and wear resistance of white cast iron.

There are also other methods used today to produce castings. In centrifugal casting, for example, molten metal is poured into a rotating mold. This method is used primarily to make tubes, pipes, and cylinders.

Some iron castings are heat treated to produce different surface characteristics for milling.

Castings should be sound and free from warp, holes, and other defects that would impair their strength or appearance. Joints should be located where they are the least conspicuous. Exposed surfaces should have a smooth finish and sharp, well-defined lines and arrises. Machined joints, where necessary, should be milled to a close fit. The necessary rabbets, lugs, and brackets should be provided so that the casting and adjacent construction can be assembled neatly.

Steel

Fairly high-quality steel was made in ancient Syria and in Toledo, in what is now Spain. The methods used were lost, however, during the Dark Ages. As a result, metals we would now classify as steel were not manufactured again until the 1730s.

So-called blister steel was made from very early times. It was the only "steel" available from the beginning of the Dark Ages until the invention of the crucible process by Benjamin Huntsman in the 1730s. Blister steel, which is not steel as we now know it, was nothing more than carburized wrought iron. The crucible process permits the melting of this carburized iron, or blister steel, into a homogeneous mass, which makes it steel in the modern sense. Such steel can be of very high quality. Unfortunately, the process is quite labor intensive and yields only small quantities of steel with each charging.

Modern steel production in quantity became possible during the nineteenth century with the invention of the hot-blast method of iron smelting, followed by the invention of the Bessemer and open-hearth processes of manufacturing steel. Electric furnaces, which permit the production of high-quality materials, such as noncorroding steel, stainless steel, high-speed tool steel, and heat-resistant steel, were introduced in the twentieth century, as was the basic oxygen process. Most steel made in this country is produced by the open-hearth process.

Steel Types

All steels are iron alloys containing 2 percent or less carbon. In fact, all iron alloys with 2 percent or less carbon are called steel, except that when the carbon content is very low the material is known as ingot iron. Steel with a carbon content of less than 0.80 percent is called hypoeutectoid steel. Steel with more than 0.80 percent carbon is called hypereutectoid steel. Each different steel, or alloy, has a different amount of carbon.

Other materials are added in controlled amounts to steel to impart special characteristics or to change the qualities of the basic material. Ferroalloys, for instance, are added to most steels. Ferromanganese is required in all steels. Other materials that may be present include phosphorus, sulfur, and silicon.

Steel is divided into four basic types: carbon, alloy, tool, and stainless.

Most structural steel, and much of the steel used in nonstructural items in buildings, is some type of carbon steel.

Alloy steels include, but are not limited to, those containing one or a combination of the following: chromium, nickel, vanadium, molybdenum, silicon, and manganese. Structural nickel and structural silicon steels have been used in bridges and other large exposed structures, but they are seldom used in ordinary buildings. Other alloy steels are used as components of elements in buildings. Chrome-nickel steel, for example, is used in making gears, axles, and other motor and machine parts for mechanical equipment, cranes, elevators, and other machines.

Tool steels are used extensively in the construction process, especially in cutting tools and drills, but are not a major element in finished buildings.

Stainless steel probably represents less than 1 percent of the total steel produced in this country, but it is used quite frequently in nonstructural metal items used in buildings. Stainless steels and heat-resisting steels are often referred to as being in the same group, because they are both classified according to their chromium content. Heat-resisting steels maintain their essential characteristics at higher temperatures than other steels. Stainless steel does not rust and corrodes only under extremely harsh conditions. The American Iron and Steel Institute (AISI) groups stainless steel into three basic categories: martensitic, ferritic, and austenitic. Heat-resistant steels are also included in the martensitic category. They have a chromium content of between 4 and 12 percent.

Stainless steels contain between 12 and 27 percent chromium. Most stainless steel used in buildings comes from the austenitic category, which contains both nickel and chromium, and sometimes manganese or molybdenum. Ferritic stainless steel is used also, however, for such items as flashings, gutters, downspouts, and in kitchen applications, such as in range hoods and appliances. Probably the most common stainless steel used in buildings is "18/8" stainless steel, so called because it contains about 18 percent chromium and 8 percent nickel.

Steel-Products Production

The essential process of steelmaking involves the melting together of pig iron, iron ore, and scrap (both iron and steel), alloying elements, and other ingredients. The other ingredients vary, depending on the process used and the type and grade of steel being made. The open-hearth process, for example, requires the addition of limestone or burnt lime to the furnace charge. The Thomas process, which is the basic Bessemer process, requires lining the furnace, called a converter, with burnt dolomite. Other processes require other ingredients. All ingredients must be carefully selected in order to produce the type of steel desired.

Regardless of the process or type of furnace used to produce the molten steel, it is poured from the furnace into ladles. From them the steel is either cast into ingot molds or poured into a reservoir called a tundish and from there is turned into a continuous ribbon of steel. Ingots are uniformly cooled in soaking pits, then used as raw material in the manufacture of the steel shapes, sheets, plates, pipes, and other products used in building and related construction. The continuous-casting process called strand casting produces a continuous ribbon of steel, which is then cut into the desired lengths. Strand casting can eliminate many of the steps necessary to produce certain products from ingots.

Forming Methods.   Steel is usable in building construction only after it has been formed into some useful shape. There are four basic methods used to produce these shapes: casting, rolling, forging, and drawing. The latter three all produce plastic changes in the steel ingots. In each of those processes shaping requires some reheating. The different shaping methods produce steel items with differing characteristics.

Stainless steel is a special case because the shape, and thus the forming method, often dictates the finish (see Chapter 4). The standard shape designations assigned to stainless steel by the AISI are plates, sheet, strip, bars, wire, pipe and tubing, and extrusions. The method of production (rolled, forged, cold finished, hot finished, and so on) and the sizes of each shape are also a part of the designation. The National Association of Architectural Metal Manufacturers's *Metal Finishes Manual for Architectural and Metal Products* contains a table showing the classifications and their characteristics.

Casting.   Intricate steel shapes are often easier to make using the casting method than any other shaping method. In addition, although steel castings are more expensive than iron castings they are usually cheaper than similar rolled or forged steel shapes. They are also tougher and stronger than similar cast- or malleable-iron castings, which often offsets the cost difference.

Steel may be cast into sand or centrifugal molds, as described earlier in this chapter for iron castings. Steel is also sometimes cast into investment molds made from high-temperature fusion materials. This technique is sometimes called the "lost-wax process," but the actual material displaced by the molten steel may also be frozen mercury or a plastic. Steel may also be cast into ceramic molds, various forms of metal molds, and graphite molds.

Castings should be sound and free from warp, holes, and other defects that impair their strength or appearance. Joints should be located where they are the least conspicuous. Exposed surfaces should have a smooth finish and sharp, well-defined lines and arrises. Where they are necessary, machined joints should be milled to a close fit. The necessary rabbets, lugs, and brackets should be provided so that the casting and adjacent construction can be assembled neatly.

Rolling.   Most carbon-steel shapes, plates, and sheets used in building and related construction are produced in sheet and strip mills. There steel ingots are rolled down to the proper thickness and size in a continuous process. During the process, reheating may occur periodically to keep the metal at a suitable working temperature.

There are many different types of rolling mills. All of them, however, use the same basic principle: the steel is passed between a series of two fixed but rotating rollers. Each set of rollers reduces the thickness of the steel in the direction perpendicular to the rollers and correspondingly increases its length. In order to produce shapes such as those of structural steel beams, columns, angles, channels, and the like (Fig. 3-4), the rollers are grooved. Welded pipe and tubes are made from rolled plate, sheet, or strips. Rods may also be made in rolling mills.

Steel may be rolled either hot or cold. Hot-rolled steel is passed through the rolling mill at a temperature above its critical working temperature. In many processes the steel is rolled hot early in the sequence, to take advantage of the plasticity of the hot metal, and cold late in the process, to produce better surfaces and permit closer control of the final dimensions of the piece. Although there may be some differences that depend on the composition of the steel, in most cases cold-rolled steel has a higher tensile strength but is less ductile than hot-rolled steel.

Forging. Originally, forging consisted of pounding the metal with a hammer on an anvil. Today the hammering of steel is done with a steam hammer, a method also called ''closed-die forging.''

Much steel shaping that would have been done in the old days by hand forging is done today with a hydraulic press in a method called power pressing. This method is also sometimes called ''open-die forging.'' Pressing usually produces a better product than does hammering. Presses can be used either for stamping out shapes or for imposing a new shape by deforming the metal over a die.

The effects of forging a metal either hot or cold are similar to those described previously under ''Rolling.''

Drawing (Extrusion). Rods may be drawn out to a smaller diameter, even to wire size, by pulling them through a circular hole in a die.

Seamless tubes and pipes can be made either by the extrusion process or by the related hot-piercing technique. There are actually several ways to accomplish the latter, but they are similar in that a hot billet is fed between pairs of spinning, tapered rollers, which forces the billet over a center punch. Tubes and pipes made in a rolling mill or by the technique

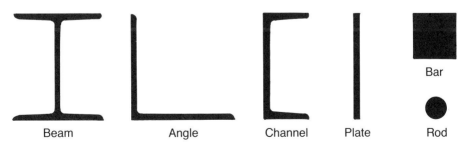

Beam   Angle   Channel   Plate   Rod

Bar

Figure 3-4 Some common shapes produced by rolling.

called seamless shaping may be stretched to reduce their wall thickness by drawing them over a centered mandrel.

Complex shapes can also be produced by extrusion. Extruded shapes are particularly used in stainless steel (Fig. 3-5).

Heat Treatment.   Steel is heat treated to change its physical properties. The term "heat treatment" applies to the process performed after the metal has been permitted to reach its initial solid state. Steel is quenched to harden it by submerging it in oil or water. Quenching, however, makes steel brittle. To eliminate brittleness it is necessary to temper the steel, by raising its temperature to a point below the critical working temperature, holding it there for a time, then quenching it to cool it again.

Working steel cold, as in a rolling mill, makes it hard and brittle. The more working of it that takes place, the harder and more brittle the steel becomes. In some cases the state reached is desirable and the steel item is used in that state. Such hardness and brittleness is not always desirable, though. Then the steel may be annealed, a process that relieves the stresses imposed during rolling, making the steel softer and easier to machine. The annealing process requires that the steel be elevated to a temperature above the critical working temperature, then permitted to cool slowly.

Steel can be made especially hard by several processes, including one known as case hardening, in which the steel is packed in charcoal and heated. This process impregnates carbon into the surface of the steel, making it harder than it would otherwise be. A less common way of hardening steel is a process called "nitriding," in which iron nitride is caused to form in the surface of the steel, making it harder than normal.

Steel Products.   Many steel products are used in producing nonstructural metal items for use in building and related construction. The following is a partial list of such products. After each item is the industry standard that that product is usually required to meet.

- Carbon-steel castings: ASTM Standard A 27.
- Steel plates, shapes, and bars: ASTM standards A 6 and A 36.
- Steel pipe: ASTM Standard A 53.

Figure 3-5   Some typical stainless steel extruded shapes.

- Cold-finished steel bars: ASTM Standard A 108.
- Stainless and heat-resisting steel plate, sheet, and strip: ASTM Standard A 167.
- Stainless and heat-resisting steel bars and shapes: ASTM Standard A 276.
- Mild-steel plates, shapes, and bars: ASTM Standard A 283. Mild steel is today's substitute for wrought iron. It is made only by the open-hearth, basic-oxygen, or electric-furnace processes. Made in four grades, it is permitted to have a lower tensile strength and yield point than does steel conforming with ASTM Standard A 36. Other requirements for the materials covered by ASTM Standard A 283 and shapes made from those materials are found in ASTM Standard A 6.
- Carbon-steel bolts: ASTM Standard A 307.
- Stainless steel pipe: ASTM Standard A 312.
- High-strength steel bolts: ASTM Standard A 325.
- Cold-rolled carbon sheet steel: ASTM Standard A 366.
- Sheet steel for porcelain enameling: ASTM Standard A 424.
- High-strength steel bolts. ASTM Standard A 490.
- Cold-formed steel tubing: ASTM Standard A 500.
- Hot-rolled steel tubing: ASTM Standard A 501.
- Wire and wire rods: ASTM Standard A 510.
- Welded stainless steel tubing: ASTM Standard A 554.
- Cold-rolled carbon and high-strength, low-alloy steel sheets: ASTM Standard A 568.
- Hot-rolled, commercial-quality steel sheet and strip: ASTM Standards A 568 and A 569.
- Hot-rolled structural steel sheet and strip: ASTM Standard A 570.
- High-strength, low-alloy steel: ASTM Standard A 572.
- Hot-rolled carbon-steel bars and bar-sized shapes: ASTM Standard A 575.
- Improved atmospheric corrosion-resistant, high-strength, low-alloy steel: ASTM Standard A 606.
- High-strength, low-alloy steel sheet and strip: ASTM Standard A 607.
- Cold-rolled structural steel sheet: ASTM Standard A 611.
- Carbon-steel bars: ASTM Standard A 663.
- Hot-wrought carbon-steel bars: ASTM Standard A 675.
- Stainless steel castings: ASTM Standard A 743.
- Steel floor plates: ASTM Standard A 786.

Notice that this list contains mostly carbon steel and stainless steel items. A few alloy steels are included, but they are seldom used to make nonstructural items used in building and related construction. Most alloy steels are high in cost, which is undesirable, and also high in strength, which is unnecessary for most nonstructural items. Typical uses for alloy steels in construction include superstructures for bridges, and very tall buildings.

The American Iron and Steel Institute (AISI) also has a designation system that identifies the composition of steel using a four- or five-digit numbering system. In addition, there is the Unified Numbering System (UNS) developed by the ASTM and the Society of Automotive Engineers. The UNS consists of a letter and five digits. Most steel used in buildings, however, is designated by using ASTM standards.

In addition to using the ASTM standards, stainless steel is often referred to by its AISI designation. Martensitic and ferritic stainless steels both fall within the 400 series of AISI designations. Type 430, for example, is often used in sheet-metal applications. Austenitic stainless steel spans the 200 and 300 series numbers. The most commonly used of these are types 302, 303, and 304. Type 302, which is the basic 18/8 grade mentioned earlier, and 304 are often listed as being interchangeable. These types are used in curtain walls, storefronts, doors, food-service cabinetry, countertops, sinks, railings, and many other uses. The 200 series stainless steels contain manganese in addition to chromium and nickel. Types 201 and 202, which are similar to types 301 and 302, are often used interchangeably with them.

Fabrication.   Formed-steel shapes, pipes, bars, and plates are used directly in fabricating many of the kinds of nonstructural items discussed in this book. Many other components, however, are made from steel sheets and strips that have been formed into shapes such as those shown in Figure 3-6. The methods used to form sheets and strips into such shapes include hot forming, cold forming, and machining. Cold forming is done by several methods, including brake forming, roller forming, stretch forming, shear forming, and deep drawing. Most sheet, strip, and tubular-steel products used to make nonstructural metal items for use in building and related construction are cold formed.

Hot Forming.   Metals that are too thick or too hard to be cold formed effectively can be formed hot. The methods of hot forming include forging, high-temperature forming, and high-energy-rate forming.

Figure 3-6   A few typical formed shapes.

Forging is essentially the same process described earlier under "Forging."

High-temperature forming, which is also called "electro forging," is roll forming in which an electric current is passed through the metal to heat it for forming.

In high-energy-rate forming, the metal is deformed by a shock wave. This wave may be generated by the detonation of small explosive charges, by creating an electric spark in a fluid, or by the sudden release of compressed gases.

Cold Brake Forming. Many sheet-metal components used in nonstructural metal items, and in shapes such as steel door frames and the like, and many roofing-related metal items are shaped by brake forming (Fig. 3-7). This is probably the most-used method for shaping sheet metal used in buildings.

In essence, brake forming consists of squeezing sheet metal between a punch and a die block. Several consecutive operations are usually required to make the final shape.

Cold Roller Forming. The usual method of producing steel decking and other corrugated-metal shapes is called "roller forming" (Fig. 3-8). It is also used to make wall panels and many other shapes. Cold roller forming is essentially the same as the roll-forming methods discussed earlier.

Cold Stretch Forming. Irregularly shaped, curved metal pieces (Fig. 3-9) are often shaped by bending sheet metal over form blocks. This method is called "stretch forming." Typically, the metal is held at its ends and pulled downward across the form block.

Cold Shear Forming. Cylinders and the conical shapes found in base flanges and caps for railing posts, and many similarly shaped metal items, are often made in a process called shear forming. In this method sheet metal is turned on a mandrel and shaped by being pressed over a form block, using external pressure produced by rollers.

A type of shear forming known as "spinning" is done by clamping sheet metal to a form block and spinning this block and the metal against a roller, which gradually forces the metal to fit the shape of the block.

Figure 3-7 A few cold-brake-formed or bent shapes.

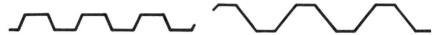

Figure 3-8 Typical roller-formed sheets.

Cold Deep Drawing.   Deep drawing is the process of forming in a draw press. This method is not used extensively to make nonstructural metal items for buildings, but some items, particularly stainless steel ones, may be produced in a draw press.

Machining.   Many items made from alloy steels and stainless steels are machined. Machining includes milling, sawing, drilling, punching, shearing, tapping, and reaming.

Aluminum

Aluminum is produced by first refining bauxite ore to produce an oxide of aluminum called alumina, which is a white, sugarlike powder. Then the oxygen is removed from the alumina in a process called "reduction." In this electrolytic process an electric current is passed through a hot (about 1,775 degrees Fahrenheit) solution of alumina dissolved in molten sodium aluminum fluoride, or cryolite. Molten aluminum then settles to the bottom of the vessel, from whence it is siphoned off into ladles. From the ladles it is poured into molds to make ingots or into furnaces for alloying.

Aluminum makes up about 8 percent of the earth's crust and is its most abundant metal. Centuries ago, the ancient Egyptians, Phoenicians, and Lydians used a form of it to set dyes. The material they used was not in the metallic form we know today, however. Theirs was an ammonia alum or aluminum ammonia sulfate.

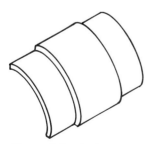

Figure 3-9 A shape that might have been cold stretch-formed.

The metal aluminum was first produced in 1825, but it was not commercially developed until after the invention in 1886 of the electrolytic process used to produce it. Today aluminum is the second most important metal used in building and related construction, following iron and its alloys (steel).

A major advantage aluminum enjoys over the iron alloys is, of course, that aluminum tends to oxidize slowly, and then only slightly. In addition, the resulting aluminum oxide tends to be nonporous, which thus protects the underlying metal from further oxidization.

Aluminum is a silver-colored metal that is very light, compared with iron and its alloys. By heat treating and alloying aluminum this metal's strength can be enhanced to give it a very high strength-to-weight ratio.

Aluminum Alloys

As it comes from the pots, aluminum is about 99.5 percent pure, and even that can be improved by further refinement, to 99.99 percent purity. Unfortunately, pure aluminum has a tensile strength of only about 7,000 psi. In addition, it is ductile and soft. To make it more usable than these characteristics permit, aluminum must be alloyed with other metals.

Alloys can be divided into two basic categories: casting and wrought. The wrought alloys can themselves be subdivided into heat-treatable and non–heat treatable alloys.

Casting Alloys. Most aluminum casting alloys contain either copper or silicon and other metals, including one or more of the following: magnesium, manganese, zinc, nickel, and iron. The simplest of casting alloys may contain up to 8 percent copper or 13 percent silicon.

Certain casting alloys may be heat treatable, but heat treatment is seldom used on cast-aluminum items, because the high strengths created by heat treatment are seldom needed in cast items.

Wrought Alloys. Wrought alloys may be either heat treatable or non-heat treatable. The alloying elements used for the first group include copper, magnesium, zinc, and silicon. The alloying elements used in the so-called common alloys, the non-heat-treatable ones, include magnesium, manganese, silicon, and iron. In each case the various alloying elements may be used either singly or in various combinations.

Heat-treatable alloys may have their strength improved dramatically by the heat-treatment processes described later in this chapter under "Heat Treatment."

Non-heat-treatable alloys do not increase in strength under heat treatment, but they may have their strength increased by cold working. This process includes rolling, strain hardening, and other cold-working methods. Series 1000 aluminum, described next, may be strengthened by strain hardening.

Alloy Designations.   The Aluminum Association has developed a method for designating the various alloys of aluminum. The designation consists of a four-digit number in which the first figure is the major alloying element, the second designates modifications of the alloy, and the third and fourth digits are arbitrary numbers that identify alloys within a series. The first number establishes the series. For example, the 1000 series contains pure aluminum, meaning 99 percent pure or higher. The alloys in this series are used to make aluminum-foil vapor retarders and in sheet-metal work.

The main alloying material in the 2000 series alloys is copper. When heat treated, some alloys in this series have properties that are equal or superior to those of some grades of steel. Structural aluminum shapes and aluminum fasteners are often made from alloys in this series. Sheets made from series 2000 alloys that are to be exposed to weather are often clad with aluminum alloys in the 6000 series, to take avantage of the superior corrosion resistance of the higher-numbered series' materials.

Series 3000 alloys use manganese as their major alloying element. Many common nonstructural items are made from Alloy 3003, including acoustical ceiling and wall panels, air-distribution ducts, doors, and sheet metals used in flashings, gutters, downspouts and the like.

Series 4000 alloys use silicon as the major alloying element. The alloys in this series are used to produce certain dark anodic finishes.

Series 5000 alloys use magnesium as their major alloying element. The aluminum alloys in this series are used to produce venetian blinds, weatherstripping, and other building items.

Series 6000 alloys contain both magnesium and silicon as major alloying elements. The aluminum alloys in this series are probably the ones most used in building and related construction. From them are made structural shapes, nails, bars, hardware, door frames, curtain walls, and storefront elements. The series 6000 alloys are heat treatable.

Series 7000 has zinc as its major alloying element. When other elements such as magnesium, chromium, and copper are added, these alloys have extremely high strengths. These are the alloys used in making aircraft parts. They are seldom used, however, in nonstructural metal items in buildings.

Temper Designations.   The temper designations consist of a letter and one or more numbers. The letter designates whether the alloy is as fabricated (F), annealed (O), strain hardened (H), solution heat treated (W), or thermally treated (T). Alloys with an H designation are non-heat treatable, those with the T designation, heat treatable.

The first number designates the type or degree of treatment. Thus, T6 means solution heat treated, then artificially aged, as explained in the following section.

The second number, which is used only with the H designation, indicates the degree of strain hardening. Thus, H24 indicates a strain-hardened, partially annealed alloy that is hardened to the 1/2 hard state.

The temper designations follow the alloy designations. Thus, a storefront section might be said to have been made from Alloy 6063-T6.

Heat Treatment

The methods of heat treating aluminum can be divided into two categories. The first, age hardening, is effective only when used on heat-treatable alloys. The second, annealing, is effective on both heat-treatable and non-heat-treatable aluminums.

Age Hardening. Heat treatment of strong alloys can raise their tensile strength to as high as 10,000 psi. The age-hardening process consists of elevating the temperature of the aluminum to a point just below its melting point in a process called "solution heat treating," then rapidly quenching it in cold water. This technique makes the aluminum very workable. It can be kept in this state by lowering its temperature further and maintaining it below freezing. If it is left to stand at room temperature, the aluminum will slowly become stronger. Some alloys will continue to become stronger for a long time.

Room-temperature aging, or natural aging, can be accelerated, with even higher strengths obtained, by heating the alloy in a process known as artificial aging.

Annealing. All aluminum alloys can be annealed. Annealing is simply elevating the alloy to a high temperature, then slowly cooling it. The process relieves strains imposed in the metal by cold working.

Aluminum Products Production

Forming Methods. Aluminum is usable in producing nonstructural metal items for use in buildings only after it has been formed into some useful shape. There are four basic methods used to produce such shapes: casting, rolling, extruding, and drawing. All but the first produce plastic changes in aluminum ingots.

Casting. Intricate shapes are often made using the casting method. Aluminum castings are made using sand molds, permanent molds, or dies.

Sand molds are the cheapest, but the shapes resulting from them are not as fine as those in the other methods, and tolerances are harder to hold.
Permanent molds consist of metal castings with metal cores. Similar molds having sand cores are called semipermanent.
Aluminum is poured into sand and permanent molds. With die-cast molds it is forced in under pressure. As a result, die castings are smoother and give more consistent results than those made in most other types of molds. Dies are used when large numbers of the same piece are required. For example, most cast-aluminum finish hardware items are made in dies.
Castings should be sound and free from warp, holes, and other defects that would impair their strength or appearance. Joints should be located where they are the least conspicuous. Exposed surfaces should have a smooth finish and sharp, well-defined lines and arrises. Machined joints should be milled to a close fit. The necessary rabbets, lugs, and brackets should be provided so that the casting and adjacent construction can be assembled neatly.

Rolling.   Many of the aluminum items used in building and related construction are, at some point in their development, run through a hot-rolling mill. Many of them are also then cold rolled to produce a harder, more consistent finish. Some of the items produced by the rolling process include plates, sheets, structural shapes, rods, bars, forging-stock rod, and redraw rods (see Fig. 3-4).
Sheet products made by rolling include coil sheets, which are either given a finish on the coil and later cut to size, or first cut to size and later finished. Other sheet products include aluminum foil and brazing sheet, which is a layer of two alloys. The alloy of the outer layer has a lower melting point than that of the core layer so that the core does not melt in the brazing process.
Rolled bars may be round, square, rectangular, or hexagonal.
Structural shapes include angles, channels, and I beams. Most structural shapes, however, are extrusions.
The rolling process consists of passing an aluminum ingot through a succession of rollers that gradually reduce its cross-section while increasing its length, until the desired shape is realized. The rollers are grooved to produce the desired shapes.

Extrusion.   Many aluminum products used in nonstructural metal items in buildings and related construction are produced by extrusion. Extruded items include structural shapes, rods, bars, seamless tubes and pipes, and special shapes used in various nonstructural metal items (Fig. 3-10). Extrusions that are large or complex are often made in more than one piece.

Figure 3-10   Extrusions may be either simple or complex.

The separate pieces are then welded together, or fastened or snapped together, in the shop or in the field, to make up the desired final shape.

Extrusions are superior in virtually every way to rolled products. Among their other superior characteristics are that they are straighter, more accurate in their dimensions, and smoother.

Drawing.   Tubes and wire can be made by drawing aluminum through a die instead of pushing it, as in the extrusion process. Redraw rod made by hot and cold rolling is used to make wire by cold drawing.

Seamless tubes and pipes, which can be made by the extrusion process, can also be made by drawing a bar or rod over a centered mandrel in the die.

Aluminum Products.   Many aluminum products are used in producing nonstructural metal items for use in building and related construction. A partial list of such products follows. After each item is the industry standard that that product is usually required to meet.

Castings: ASTM standards B 26 or B 108.

Sheet: ASTM Standard B 209.

Plate: ASTM Standard B 209.

Drawn tubes: ASTM standards B 210 or B 483.

Bars, rods, and wire: ASTM Standard B 211.

Extrusions: ASTM standards B 221 or B 429.

Forgings: ASTM Standard B 247.

Cladding.   Some alloys of aluminum have superior corrosion resistance; others have higher strength. To take advantage of both these characteristics, sheets of the stronger alloys are clad with sheets of the more corrosion-resistant ones. The cladding material is metallurgically bonded to the core material. This product class is called Alclad.

Most clad products are in sheet form, but other shapes, such as tubes and rods, are also available. For example, Alclad 3004 is a commonly used sheet material. Alclad 3003, which is used for extruded bars, rods, wire, shapes, and tubes, is made up of a 3003 alloy material that has been clad on both sides with a 7072 alloy material.

Fabrication.   Some rolled, extruded, and drawn products are used exactly as they come from the mill to fabricate the kinds of nonstructural items discussed in this book (see Chapter 6). Others are used in the shape in which they come from the mill, but with additional finishing (see Chapter 4). Many other components, however, are made by fabricating those products into shapes like those shown in Figures 3-6 through 3-9. The methods used to fabricate such products into the kinds of shapes shown include additional roll forming, forging, bending, stretch forming, and machining.

Roll Forming.   The rolled products mentioned earlier are passed through an additional roll-forming process to make the final product. The process is the same as that described earlier, except that most post–mill rolling is cold rolling.

Forging.   Forging is done using one of two methods, each of which uses closed steel dies. In the first method, hammering, the aluminum is hammered into a die cavity by repeated blows of a drop hammer. The second method, pressing, involves forcing the metal into the die cavity under hydraulic pressure.

Both methods result in aluminum that is strong and has high fatigue resistance. These characteristics can be further improved by heat treatment.

Aluminum finish hardware items are often produced by forging.

Bending.   Many aluminum sheet-metal and plate components used in nonstructural metal items, and shapes such as door frames and the like, as well as many roofing-related aluminum items are shaped by bending (see Figs. 3-6 through 3-8).

Some alloys can be bent to a sharp 90-degree angle or more, but others resist such severe bending. Bending is often used to create closers and corners in storefronts and curtain walls. A well-made bent section can simulate extruded shapes, but it may be much cheaper than an extruded shape for a one-time use.

Stretch Forming.   Irregularly shaped curved aluminum pieces are sometimes shaped by stretching an aluminum sheet or plate while bending it over one or more form blocks (see Fig. 3-9). This method is called "stretch forming." Typically, the metal is held at its ends and pulled downward across a form block.

Machining. Many items made from aluminum are machined, often after they have been cast, rolled, or extruded into the desired shape. Machining includes sawing, drilling, punching, shearing, tapping, and reaming.

Copper Alloys

When all uses are considered, copper is probably the world's second most used metal, next to the various alloys of iron, including steel. Copper appears in nature in its metallic state, which probably explains its use as early as 5000 B.C. Today copper is obtained mostly by refining the ores cuprite, which is copper oxide; malachite, which is basic copper carbonate; or chalcocite or chalcopyrite, both of which are copper sulfide. The refining process is a complicated one requiring roasting, smelting, and converting of the ore, then refining the resulting metal by electrolysis.

Newly refined copper is about 99.95 percent pure. It has a reddish-yellow color and is soft, malleable, ductile, relatively strong, and a very good conductor of electricity. Copper does not corrode in dry air, but it will oxidize when heated. In normal atmosphere it oxidizes slowly, forming a thin surface coating of verdigris, a material that is similar to the ore malachite.

Copper is used for many purposes, including the production of copper sulfate, which is used in printing, electroplating, and to purify public water supplies. It is also used in tanning leather and in manufacturing pigments, insecticides, dyes for cotton, and even as an astringent. Copper is also widely used in coins, cooking utensils, wire, art objects, and various sheet-metal items such as gutters, downspouts, and roofing. One of the primary uses of copper is in making various alloys, including the two most interesting to those in the building industry, brass and bronze.

The Copper Alloys

There are more than 1,100 recognized alloys of copper. Of these, more than half are alloys of copper and zinc or copper and tin. The first has traditionally been known as brass, the second as bronze. Today most copper alloys used for other than utilitarian purposes in building and related construction are called bronze, even though the many that contain zinc are really brass.

Bronze was almost certainly the first industrial alloy. In some areas of the world copper and tin ore are found naturally intermixed, which probably explains the appearance of bronze as early as 3500 B.C. Brass followed much later, but even it has been in use for other than artistic purposes since about 50 B.C.

Alloy Designations.   The Copper Development Association (CDA) has a numbering system for copper alloys. The CDA is also the administrator of the Unified Numbering System (UNS) mentioned earlier, which was developed jointly by the ASTM and the Society of Automotive Engineers. The CDA system consists of three digits. The UNS adds a letter C prefix to these three digits, and a 00 suffix. For example, the alloy 280 (Muntz metal) in the CDA system is C28000 in UNS notation.

Common Alloys.   The CDA organizes the most common copper alloys used in the construction industry into three groups. The A group contains copper, including Alloy 110, which is 99.9 percent pure copper, and Alloy 122, which is 0.02 percent phosphorus.

The B group contains the common brasses and extruded architectural bronze. It includes Alloy 220 (commerical brass), which is 10 percent zinc; Alloy 230 (red brass), 15 percent zinc; Alloy 260 (cartridge brass), 30 percent zinc; Alloy 280 (Muntz metal), 40 percent zinc; and Alloy 385 (architectural bronze), which is 40 percent zinc and 3 percent lead.

The C group contains the so-called white bronzes. It includes Alloy 651 (low-silicon bronze), which has between 1.5 and 3 percent silicon; Alloy 655 (high-silicon bronze), 3 percent silicon; Alloy 745 (nickel-silver), 25 percent zinc and 10 percent nickel; and Alloy 796 (leaded nickel-silver), which has 42 percent zinc, 10 percent nickel, 2 percent manganese and 1 percent lead.

Casting alloys are special alloys made specifically for casting. They fall in the 800 and 900 series of CDA alloy designations.

Other alloys are used also, of course, as when color matching is required. The CDA booklet *Copper, Brass, Bronze Design Handbook: Architectural Applications* includes a chart showing which alloys match each other in color. When a nonstructural metal item contains several different shapes or types of components, such as sheets, pipes, extrusions, fasteners, or castings, the different parts are most likely made from different alloys, even though their colors match. For example, possible matches for the color of Alloy 230 (red brass) sheets include Alloy 385 (architectural bronze) extrusions, Alloy 836 (a casting alloy) castings, Alloy 280 (Muntz metal) fasteners, and Alloy 655 (high-silicon bronze) filler metals.

Copper-Alloy Products Production

Forming Methods.   Copper alloys are usable in producing nonstructural metal items for use in buildings only after they have been formed and fabricated into some useful shape. There are four basic methods generally used to work the shapes later fabricated into the final forms used to produce nonstructural metal items. These techniques include casting, rolling, extruding, and drawing.

Casting.   Intricate shapes are often made using the casting method. Castings should be sound and free from warp, holes, and other defects that impair strength or appearance. Joints should be located where they are the least conspicuous. Exposed surfaces should have a smooth finish and sharp, well-defined lines and arrises. Machined joints should be milled to a close fit. The necessary rabbets, lugs, and brackets should be provided so that the casting and adjacent construction can be assembled neatly.

Rolling.   Many of the copper-alloy items used in building and related construction are at some point in their development run through a hot-rolling mill. Many of them are then passed through a cold-rolling process to make their surfaces brighter and smoother. Items produced by rolling include plates, sheets, shapes, rods, bars, and forging stock.

Extruding.   Many copper-alloy products used in nonstructural metal items in building and related construction are produced by extrusion. The range of extruded items includes structural shapes, rods, bars, seamless tubes and pipes, and the kinds of special shapes often seen in nonstructural metal items (Fig. 3-11).

Extrusions are superior in virtually every way to rolled products. For example, they are straighter, more accurate in dimension, and smoother. They are, however, limited to maximum diagonal cross-sections of 6 inches and thicknesses of about 1/8 inch.

Drawing.   Tubes and wire can be made by drawing copper alloy through a die, instead of pushing it, as in the extrusion process.

Laminating.   Copper-alloy sheets and strips are adhesively bonded under pressure to steel, aluminum, plywood, and other materials in a process called laminating.

Figure 3-11   Some typical copper-alloy extruded shapes.

Copper-Alloy Products. Many copper-alloy products are used in producing nonstructural metal items for use in building and related construction. A partial list of such products follows. After each item is the industry standard that that product is usually required to meet. For the products not followed by a standard there is no currently recognized industry standard known to the author. Following the industry-standard reference is the CDA alloy number and name for the material. The alloys named are representative of only a single group of materials that generally match in color and are by no means the only alloys used for the product listed. Refer to the CDA publication *Copper, Brass, Bronze Design Handbook: Architectural Applications* for an indication of other alloys used for the purposes listed.

Extrusions: ASTM B 455; Alloy 385, architectural bronze.

Sheets, plates, and bars: ASTM B 370; Alloy 280, Muntz metal.

Round tubing: ASTM B 135; Alloy 230, red brass.

Castings: ASTM B 271, Alloy 857. For sand castings: ASTM B 584.

Wire and rods: Alloy 280, Muntz metal.

Fabrication. Some rolled, extruded, and drawn products are used exactly as they come from the mill to fabricate the kinds of nonstructural items discussed in this book (see Chapter 6). Others are used as they come from the mill but with additional finishing (see Chapter 4). However, many other components are made by fabricating those products into shapes such as those shown in Figures 3-6 through 3-9. The methods used to fabricate products into the kinds of shapes shown include additional roll forming, forging, bending, brake forming, explosive forming, and spinning. Not all methods are applicable to all alloys. Refer to the CDA publication *Copper, Brass, Bronze, Design Handbook: Architectural Applications,* which indicates the alloys that can be formed by each method.

Roll Forming. In this technique the rolled products mentioned earlier are passed through an additional roll-forming process to make the final product. At this stage, roll forming consists of passing sheets through a series of shaped rollers to produce the desired configurations.

Forging. Forging is done by one of four methods. The first two each use closed steel dies. In the first method, hammering, the copper alloy is hammered into a die cavity by repeated blows of a drop hammer.

The second method, pressing, involves forcing the metal into the die cavity under hydraulic pressure.

In the third process, hydroforming, a punch or formed-steel die is forced, under hydraulic pressure, against the metal, which is backed up by rubber.

In the fourth method, stamping, the metal is bent, shaped, cut, indented, embossed, coined, or formed by means of a press or hammer that is faced with a shaped die.

Bending.   Many copper-alloy sheet metal and plate components used in nonstructural metal items, and shapes such as door frames and the like, and many roofing-related copper items are shaped by bending.

Brake Forming.   This process is used to shape many copper-alloy components used in nonstructural metal items, shapes such as door frames and the like, and many roofing-related copper-alloy items. In fact, brake forming is probably the most-used method for shaping copper-alloy sheets used in buildings.

In essence, brake forming consists of squeezing sheet metal between a punch and a die block. Several consecutive operations are usually required to make a final shape.

Explosive Forming.   High-energy-rate forming is done by forcing metal against a single die, using a shock wave generated by detonating small explosive charges.

Spinning.   A type of shear forming called "spinning" is done by clamping sheet metal to a form block and spinning the block and metal against a roller, a process that gradually forces the metal to fit the shape of the form block. Spin forming is done using either power or hand tools.

Other Metals

There are several other metals commonly used in nonstructural items for building and related construction.

Zinc

Zinc is used occasionally in nonstructural items where strength is not paramount. Most zinc is used, however, as an alloying metal or a coating for other metals. Refer to Chapter 4 for a discussion about zinc coatings, such as the galvanizing of iron and steel.

Zinc is a bluish-white metal that is plastic and highly corrosion resistant. It is not, however, decorative, and is thus usually given an applied finish, mostly paint and other organic coatings, whether it is used pure or as a coating over another metal.

Another major use of zinc is as an alloying metal. It is used with copper to make brass and with aluminum to produce high-strength heat-treatable alloys. Zinc is also used in magnesium–aluminum alloys to increase their ductility.

Zinc is also used as a pigment in some paints (see Chapter 4). It is particularly useful in primer paints for galvanized surfaces and on rusted steel surfaces.

Lead

Lead, a heavy metal, is very ductile, which makes it soft and weak. But this very ductility, as well as its resistance to corrosion and its density cause it to be used extensively in the construction industry.

Lead can be easily rolled into sheets and extruded to make pipes. It is used extensively in the chemical industry because of its resistance to corrosion by many chemicals, including sulfuric acid. Its chief use in buildings is in sheet form as radiation protection, sound barriers and baffles, and to coat other metals for corrosion protection. Lead-coated copper, for example, is a common roofing material.

Lead is also used in paints, especially in primer coatings on steel. This use may diminish in the future because of fears of lead poisoning, especially in children, but lead paints will continue to be found on existing metals for many years, even if their use is eventually banned altogether.

Chromium

Chromium, which is very hard and highly corrosion-resistant, is seldom used alone. Its largest uses are as a coating on steel or other metals and as an alloying metal for steel. All stainless steel, for example, contains chromium.

Chromium is so hard that it resists polishing. Therefore, to produce the common highly polished chrome-plated surfaces chromium must be deposited over a highly polished steel surface, usually with an intermediate layer of nickel.

Magnesium

The chief use of magnesium in the building industry is as an alloying metal for aluminum and steel.

Nickel

Nickel, a silver-white metal highly resistant to corrosion, is used as a finish coating on some nonstructural items used in buildings. Its primary use in buildings, however, is as an alloying metal. Monel metal, for example, is an alloy of nickel with 31 percent copper. Nickel silver is an alloy of nickel and brass. Perhaps half the nickel produced in this country is used in making steel alloys.

At one time nickel was a popular finish coating for steel. Today it is more likely to be used as an undercoating over steel below the more popular chromium.

Tin

Probably the best-known use of tin in the building industry today is as an alloying metal with copper to make bronze. Tin is also used as a coating for steel and iron (see Chapter 4). In earlier days much tinplate was used in roofing. Alloys of tin and other metals such as zinc are used for coatings on steel. Solder contains tin (see Chapter 6).

Tin is used less than zinc as a protective coating for steel, because tin is much rarer than zinc, especially in North America, and because tin does not protect steel as well as zinc does, for reasons explained in Chapter 4 under ''Metallic Finishes'' and Figure 4-1.

Gold

The precious metal gold is seldom used in the building industry except to gild decorative items. It is worth mentioning here, however, because there are many buildings, especially older ones, in which gold leaf was used extensively. Gold for gilding is not quite as expensive as it might appear, because of this metal's great malleability—gold leaf is only 1/250,000 of an inch thick. At this thickness an ounce of gold will cover 120 square feet of surface.

Metal Material Failures and What to Do about Them

There are many types of failure that might occur in nonstructural metal items used in building and related construction and their finishes. Many such failures are discussed in Chapters 4 through 7. There are, however, only two basic types of failure experienced by the actual metals themselves: fracture and corrosion. Each may result from a number of different causes.

One major problem that may contribute to either corrosion or fracture or both is poor material selection. Most often, such failure results from the wrong metal being selected for the particular metal item and location.

Another major reason for failure is not protecting susceptible metals from corrosion.

Types of Corrosion

As generally defined, "corrosion" means the wearing away of a material, usually gradually. Here, however, we will broaden the definition to include processes that change a useful metal into a useless chemical compound, thus destroying its desirable characteristics or properties, and processes that erode the metal.

Most of the corrosion we need to worry about when dealing with nonstructural metal items affects only the surface of the metal. The other types of corrosion that affect the interior structure of the metal (structural corrosion) may become a factor in some instances, however. Corrosion weakens a metal and causes pits and crevices. In a beginning fracture corrosion can contribute heavily to continuing fracture failure. Fracture and its corrosion components are discussed under "Types of Fracture" later in this chapter.

If left unprotected, iron—and most alloys containing iron—will corrode rapidly. The alloying of corrosion-resistant metals with metals that corrode readily helps the resultant alloy resist corrosion. Thus, stainless steel is corrosion resistant because it contains highly corrosion-resistant chromium. Most of the nonferrous metals resist corrosion (especially oxidation) to some extent, but no metal is truly corrosionproof.

The types of corrosion that affect the kinds of metals used in nonstructural items in building and related construction include oxidation, erosion due to weathering and exposure, chemical attack, galvanic corrosion (electrolysis), and structural corrosion.

Oxidation.   The most common form of corrosion that damages ferrous metals is the oxidation of the metal, which forms ferric or ferrous oxide. When water, either as a vapor or a liquid, contacts the ferric or ferrous oxide, hydration occurs. The resulting formation may be one of several possible compounds, including hydrated ferric oxide, ferric oxyhydroxide, and hydroxide—all commonly called rust. When chlorides, carbonates, sulfur, or other chemicals are present, rust may include corrosive products such as ferric chloride, iron sulfide, or ferric oxychloride. Rust may be yellow, bright orange, dark reddish brown, red, or black. A ferric metal that has been buried in the ground may contain ferrous chloride tetrahydrate, which is corrosive. When it comes into contact with moist air a process known as "sweating" may occur in which droplets of hydrochloric acid appear on the metal. The presence of salt, as in seawater or salt sea air, will enhance the process of oxidization and increase its severity and the speed with which it occurs.

Some ferrous metals are more likely to oxidize than others. Wrought iron, for example, is less likely to rust than its modern substitute, mild steel. Cast iron will usually form a light scale, then resist further rusting. Casting flaws in cast iron can mimic corrosion, which sometimes leads people to believe that the material is subject to more corrosion than it actually is. On the other hand, those very flaws may eventually become the sites of corrosion.

Nonferrous metals also oxidize, but the oxidation products that form on most of these metals that are used in building construction are not destructive the way those are that form on iron and its alloys. The green oxidation that forms on bronze, for example, not only protects the metal from further corrosion but also gives a desirable coloration to the metal. The green material is actually an alkaline oxide of the copper component in the bronze.

The oxide that forms on aluminum will not necessarily protect the metal in severe environments, but under normal conditions it will prevent further oxidation. In addition, the coating is unobtrusive in appearance and only about one ten-millionth of an inch thick. Anodizing, which is probably today's favorite method for finishing aluminum, amounts in principle to nothing more than increasing the thickness and effectiveness of aluminum's natural oxide film. In fact, the term "anodic oxidation" implies that the oxide film was created deliberately. Refer to Chapter 4 for a discussion of anodic finishes.

Weathering and Exposure.   Wind-driven materials such as sand, dirt, and hail can erode the surface of metals. Ice crystals may also damage some surfaces. However, much of the damage that people assume to have been caused by wind-driven debris is actually caused by chemical attack.

Chemical Attack.   Both ferrous and nonferrous metals are subject to damage by both liquid and airborne chemicals. These metals are attacked by acids, alkalies, carbonates, fluorides, and other chemicals used in the construction and maintenance of buildings. Acids used to clean masonry are a particularly odious problem. Wet alkaline materials such as mortar, plaster, and concrete are a problem. Chlorides and sulfates in soil will also attack metals severely.

Metals in industrial areas, where the air is likely to contain more pollutants than elsewhere, tend to corrode more quickly than those in rural areas. Metals will also corrode more quickly in seacoast areas. In salt air, cast iron may corrode so badly that it collapses in on itself. This effect is intensified in southern climates, where the sun is hotter.

The effects of pollution, such as from acid rain, while a major problem for most metals are more pronounced on some than on others. Wrought iron is affected more by pollution than are some other ferrous metals, for example. When it is subjected to severe environments, wrought iron may actually delaminate. Cast iron, which is stable in more benign environments, may completely disintegrate when subjected to high concentrations of acid rain. When acid rain spreads over a large area of copper, such as a roof, it forms a patina or basic copper sulfate that may not present the desired appearance but may actually protect the copper from further corrosion. On the other hand, copper may corrode completely if an accumulation of acid rain is permitted to stand on it as might happen in a copper gutter.

Unfortunately, there is no general agreement in the construction industry about the effects on metals of the chemicals in some building products. Some guides say, for example, that the tannic acid in red cedar will adversely affect galvanized steel, but others disagree. When there seems to be doubt about whether the components in a given material will be harmful to a metal material, the best bet is to assume the worst and protect the metal. When corrosion has already occurred and there is doubt about its cause, it is a good idea to verify whether an adjacent material is at fault.

Metals can be protected from chemical attack by preventing their contact with the offending material. One way to accomplish this is to separate the metal physically from the offending material. When this is not practicable, the metal can be coated with a nonreactive material to protect it. Some metal surfaces that will be in contact with concrete or masonry, for example, can be covered with an alkali-resistant coating such as heavy-bodied aluminum paint. Care must be taken to ensure that the material used to separate the metal in question from an offending material does not itself harm the metal.

Galvanic Corrosion.   Most metal corrosion results from an electrochemical process. Galvanic corrosion, or electrolysis, occurs when dissimilar metals are each contacted by an electrolyte such as water. This process is similar to that which takes place inside an automobile battery, where one metal acts as a cathode, the other an anode. The electrolyte causes an electric current to flow between the two metals. The anode metal dissolves and hydrogen ions accumulate on the cathode metal.

It is not always necessary to have two different metals present for galvanic corrosion to occur. Many metal alloys, such as steel, for example, contain anodic and cathodic areas. In the presence of an electrolyte an electrochemical reaction can occur between different areas in the same metal. The presence of liquid water is not necessary—even moderately high humidity can provide enough water for the action to occur. Preventing galvanic corrosion requires not allowing an electrolyte to contact both parts of the metal.

The electrolyte can be rain water or just condensation. The effect is more severe when dissimilar materials are close together, but even the wash off from one material can corrode a dissimilar one. Water washing across copper, for example, will severely corrode an aluminum or zinc surface. The same water may also cause steel to corrode, but the reaction will probably not be as severe, depending on the alloy of the steel.

Some combinations of materials are more susceptible to galvanic corrosion than are others. Galvanic tables have been developed that show which materials are likely to create galvanic corrosion. The propensity of given materials for galvanic action depends on where the two materials fall on the galvanic table. The farther apart the two are on the table, the more severe will be the corrosion. For example, aluminum, which falls near one end of most galvanic tables, will corrode when allowed to become wet while in contact with lead, brass, bronze, iron, or steel. When dissimilar materials are attacked by galvanic action, it is the material that is highest in the galvanic table that will corrode.

The following list of metal materials includes many that are commonly used in making nonstructural metal items for use in building and related construction. The list is arranged roughly in galvanic-table order. The higher-listed metals are anodic (electronegative) and tend to corrode when an electrolyte connects them to a metal lower on the list, which is cathodic, or electropositive.

Zinc	Tin
Aluminum	Terne
Aluminized steel	Lead
Cadmium	Nickel
Galvanized steel	Brass
Chromium	Bronze
Iron	Copper
Cast iron	Silver
Steel	Gold
Stainless steel	

The order of the metals in this list should be viewed with some skepticism, because it is only an approximation gleaned from the available references, which do not entirely agree with each other. Neither do sources that do not include such tables agree about which materials are compatible. Inconsistent data should be taken as somewhat suspect when trying to determine causes of existing corrosion. Perhaps someday some organization will study galvanic action in detail sufficient to resolve the conflicts that now exist. Meanwhile, probably the best way to resolve a corrosion problem that persists after other possible causes have been eliminated is to assume that the culprit is an adjacent material, regardless of what a published galvanic-series table or other source says.

There are two ways to prevent corrosion due to galvanic action. One is to avoid using dissimilar metals in close contact. When this is not possible it is necessary to prevent an electrolyte from contacting both metals. A heavy-bodied bituminous or another paint may sometimes be used to coat one of the surfaces. For instance, aluminum and the copper alloys may be painted with zinc-chromate paint or a heavy-bodied bituminous paint. Aluminum should not be coated with tar or creosote, however, because these substances contain acids that are harmful to aluminum. Metals can also be separated with moisture-resistant building felts or other means.

Structural Corrosion.   The type of corrosion that takes place within a metal, between its metal crystals or in pores created during the forming or fabricating process, is more insidious than the corrosion types previously discussed, because often it cannot be seen until it is too far advanced to do anything about it. Failure due to this type of corrosion is usually, but not always, accompanied by applied stress, particularly fatigue stress.

Stress corrosion can occur in any metal, but it is more common in the copper alloys and in some aluminum alloys. Pure metals seldom have this problem.

The beginning of stress corrosion occurs during cold fabrication. In the copper alloys, annealing will often help prevent failure from stress corrosion by helping relieve some of the internal stresses that contribute to stress corrosion failures. In aluminum the previously discussed process known as age hardening will sometimes set up thin layers of corrosion within the metal. Correct heat treatment can remove this problem. And protecting the metal with a corrosion-resistant coating can prevent it from getting worse.

Errors That Lead to Corrosion

All forms of corrosion are not directly related to the presence of water, but many are, and others occur more rapidly or become more severe in the presence of water. Water may contact unprotected metal because of the types of problems discussed under "Other Building Element Problems" in Chapter 2. Although such problems may not be the most probable causes of nonstructural metal corrosion, they can be more serious and costly to fix than the types of problems discussed in this chapter. Consequently, the possibility that they may be responsible for a nonstructural metal corrosion failure should be investigated.

The other building element problems discussed in Chapter 2 should also be ruled out or rectified, if found to be at fault, before repair of corroded metal is attempted or a new nonstructural metal item installed. It will do no good to repair existing or install new nonstructural metal items when a failed, uncorrected, or unaccounted-for problem of the types discussed in "Other Building Element Problems" in Chapter 2 exists. Under these circumstances the repaired or new installation will also fail. Once the problems discussed in Chapter 2 have been investigated and found to be not present, or have been repaired if found, the next step is to discover any additional causes for the failure and correct them.

There are several types of errors that will, if committed, almost certainly lead to corrosion of one type or another. The possibilities include those discussed in the following paragraphs. The outward appearances, or evidence, of corrosion failures in nonstructural metal items are listed and discussed in Chapters 4, 5, and 7 under "Evidence of Failure."

The most prevalent error to result in metal corrosion in nonstructural metal items is selecting the wrong material for an installation. One example is using a ferrous material in a highly corrosive atmosphere, such as inside a structure that houses a swimming pool. As discussed in Chapter 4, metals, even when protected, may corrode when corrosion products are present.

The second most frequently committed error that causes such metal to corrode is failing to protect the metal. When this occurs, any one of the following may be at fault:

- Failing to coat or galvanize ferrous metals so that they do not rust.
- Failing to protect ferrous and nonferrous metals alike from chemical attack.
- Assuming that galvanized metal is immune from corrosion and thus failing to protect it. Permitting galvanized metal to come into contact with reactive materials will cause corrosion. For example, the soluble sulfites in cinders will rapidly corrode zinc. The zinc coating on galvanized metal that will contact people, animals, or moving equipment or machinery will eventually wear off and the metal beneath will corrode.
- Failing to separate ferrous and nonferrous metals from mortar, plaster, concrete, and other harmful materials that might cause corrosion.
- Failing to separate one metal from dissimilar metals to prevent galvanic corrosion.
- Failing to renew and repair protection when damage occurs.

Types of Fracture

The second major type of failure that occurs in metals besides corrosion is fracture. Eventually, even many corrosion failures show up as cracks.

There are several reasons that metals used in nonstructural metal items in building and related construction might crack. Before we discuss these potential causes of failure, however, it will be helpful to know a little about the types of fracture failure that can occur in metals. There are basically four types of fracture: stress-corrosion fracture, ductile fracture, brittle fracture, and fatigue fracture.

Stress-Corrosion Fracture.   Fracture failure may be directly related to corrosion, which starts the failure by enlarging existing pits and crevices in the metal or causing new ones. Metal, especially when it has been cold worked, has internal stresses created in the forming process and may have local inhomogeneities that result in anodic-cathodic adjacent areas. Those stresses tend to congregate at the locations of the various pits and crevices. The combination results in cracks in the metal, beginning at those locations. Corrosion-weakened areas and their resulting cracks can also occur at inhomogeneities.

Ductile Fracture.   When a ductile metal is loaded above its elastic limit, it fails. The excessive stress may be tensile (involving tension), compressive (involving compression), or shear (tangential to the section) stress.

Brittle Fracture.   When a metal with limited ductility such as cast iron is subjected to a sharp blow, cracks in the metal develop, beginning at the site of the initial stress.

Fatigue Fracture.   Cyclic stressing over a long period of time will cause metals to fracture. Cracks then occur, usually starting at the locations of imperfections in the metal or at sites with corrosion.

Metal Identification by Fracture Characteristics

It is sometimes possible to identify a fractured metal by its characteristics within the fracture, though such identification is usually done in a laboratory. The nick-bend test has long been used to distinguish wrought iron from steel, for example. This test requires nicking a piece of metal and then bending it to the point of failure. Steel will break fairly cleanly, but heat-treated high-strength alloy steel will leave a serrated edge. Wrought iron will disclose a fibrous structure and not break cleanly.

The texture of the metal in a break often gives away the type of the metal. Ductile metals usually reduce in cross section, or neck down, near a fracture. Thus, an iron-alloy metal that breaks cleanly and does not neck down is probably gray cast iron or high-carbon steel, which are nonductile. A silky texture in broken ferrous metal suggests that it is low-carbon steel, because high-carbon steel will exhibit a fine crystalline appearance. Cast iron will appear either finely or coarsely crystalline.

Errors That Lead to Fracture

Since by definition nonstructural metal items are supported by something else, the root cause of fracture failure in metal that has been made into a nonstructural metal item may be traced to a failure in the metal's supports. Potential supports are discussed in Chapter 2. The causes of support-related metal failure discussed there include concrete or steel structure failure, concrete or steel structure movement, metal framing and furring problems, wood framing and furring problems, and solid substrate problems. Those possible problems, though perhaps not the most probable causes of non-structural metal fracture failure, are more serious and costly to fix than the types of problems discussed below. Consequently, the possibility that one of them may be responsible for a nonstructural metal failure should be investigated.

The causes of failure discussed in Chapter 2 should be ruled out or rectified, if found to be at fault, before nonstructural metal repairs are attempted or a new metal item installed. It will do no good to repair an existing nonstructural metal item or install a new one when a failed, un-corrected, or unaccounted-for problem of the types discussed in Chapter 2 still exists. The new installation will also fail. After the problems discussed in Chapter 2 have been investigated and found not to be present or have been repaired if found, the next step is to discover any additional causes for the failure and correct them.

The following discussion covers several types of causes of failure that if committed may lead to fracture failure. The outward appearances or evidence of fracture failure in nonstructural metal items are listed and discussed in Chapter 7 under "Evidence of Failure."

The most frequently committed error leading to metal fracture in non-structural metal items is that of selecting the wrong material for an instal-lation. Using a ductile material where it will be loaded beyond its elastic limit is one example. A less obvious one is to use ferrous metals, especially sheets and other thin sections, in a highly corrosive atmosphere, such as inside a swimming-pool equipment room, where the air is often laden with chlorine and other chemicals that are caustic to ferrous metal and where even liquid chemicals may come into contact with the metal. The problem is exacerbated when the metal is loaded and the load is then removed cyclically so that it bends back and forth. Even when it is seemingly pro-tected, as discussed in Chapters 4 and 5, a metal may corrode in such an environment. Finishes are seldom perfect. Even finishes that completely cover the metal may become scratched or chipped, leaving the bare metal open to attack. Corrosion along fracture lines also contributes to fracture failure and hastens its progress after fracture has begun.

A much less frequent cause of fracture is failing to manufacture a metal properly and in accordance with recognized standards. The designer has a right to expect that a metal material will conform to the standards claimed and have the properties required by that standard. Improper heat treatment may leave a metal too brittle, for example. An improperly made casting may contain imperfections or impurities that will cause the metal to fracture. Or a variation from the correct proportion of materials in a steel alloy might produce steel with properties quite different from those expected.

Subjecting a metal to stresses that it is incapable of supporting safely — in effect violating the designer's intention for the item fabricated from the metal—may cause the metal to buckle and fracture. Loading a metal item in ways not intended in the original design can result in imposing stresses on it that exceed its elastic limit, for example, resulting in permanent deformation. Failing to protect a ferrous metal from imposed stresses that the material cannot safely support is a similar problem. The effect is the same, whether the excess stress is imposed deliberately or unknowingly.

Repairing Corroded or Fractured Metals

Corroded metals can often be cleaned and protected, as discussed in Chapters 4 and 5. However, metals corroded beyond repair must, of course, be removed and new materials provided. The new materials should be protected properly to prevent their failing for the same reason the replaced item failed.

Fractured metals are often not directly repairable, because fractures often involve more damage than is apparent. For instance, the grain structure of a material may be damaged at a considerable distance beyond the actual crack. Sometimes a strengthening plate can be added to support a fractured metal item, but such repairs are often not satisfactory. The fractured component must usually be removed and a new component substituted. The new component may have to be of a different material or thickness to prevent a recurrence of the fracture.

Before any metal item is repaired, the advice of its fabricator should be sought out and followed. Where there is doubt about accepting the fabricator's advice, professional help should be sought (see Chapter 1). Particularly when the failure is one of fracture the prevention of future failure requires that repairs not be undertaken without careful, knowledgeable investigation and the application of professional expertise and judgment. The repair should be carried out by experienced workers under competent supervision.

Where to Get More Information

The Sheldon W. Dean and T. S. Lee–edited *Degradation of Metals in the Atmosphere*, published in 1988, heads a list of ASTM books on the subject of metals corrosion. It includes articles on the corrosion performance of weathering steel and other structural metals, as well as articles about the corrosion of stainless steel, architectural copper, wrought-aluminum alloys, zinc, galvanized steel, and other metal materials. It includes discussions of materials performance, environmental considerations, and test methods. Other ASTM books cover related aspects of these subjects in addition to methods of testing and monitoring corrosion by nondestructive and other methods. Anyone concerned with the corrosion of existing metals or the prevention of such corrosion may want to contact ASTM and ask for a list and description of ASTM books on the subject.

Harold B. Olin, John L. Schmidt, and Walter H. Lewis's 1983 edition of *Construction Principles, Materials, and Methods* is a good source of background data about iron, steel, and aluminum production and finishing. Unfortunately, it contains little that will help specifically with troubleshooting failed metals.

The Copper Development Association (CDA) has many valuable publications that will help with understanding copper alloys. The *Copper, Brass, Bronze Handbook: Sheet Copper Applications* is of particular help regarding the corrosion of copper alloys and its prevention. This work also discusses the corrosion of other metals by contact with copper and its prevention. Other useful CDA publications are marked in the Bibliography with a [3].

Until recently, the Zinc Institute was a source for data about zinc in the construction industry. However, a recent call to the number listed in the Appendix netted a recorded message that the Zinc Institute was no longer in operation and questions should be addressed to zinc suppliers.

Some AIA *Masterspec* sections contain data that may be useful on the subjects covered in this chapter. Unfortunately, however, they do not discuss damage or repairs to metal. Those sections that may be helpful are marked with a [3] in the Bibliography.

Mario J. Catani's 1985 article "Protection of Embedded Steel in Masonry" contains a great deal of valuable information about why steel corrodes and how to prevent it. Although the article is specifically aimed at embedded steel, the principles in it apply to all steel corrosion.

J. Scott Howell's 1987 article "Architectural Cast Iron: Design and Restoration" contains a table showing the various types of alloys generally used in iron, steel, stainless steel, aluminum, and copper alloys.

Jack C. Rich's book *The Materials and Methods of Sculpture* contains a great deal of relative data about metals, especially on castings. It discusses the physical characteristics of various metals and their finishing, and includes data about cleaning and retouching bronzes and other copper alloys.

Ramsey/Sleeper's *Architectural Graphic Standards* has extensive information about steel and aluminum shapes, bars, and wire, including information about their shapes, sizes, and thicknesses.

Also note other entries in the Bibliography that are marked with a [3].

Mechanical, Chemical, and Inorganic Metal Finishes

The finishes used on nonstructural metal items used in building and related construction are divided in the industry into three broad categories: mechanical finishes, chemical finishes, and coatings.

Mechanical finishes include finishes left by the metal-production process and finishes made by mechanically altering a metal's surface by such actions as grinding, polishing, and rolling. They do not involve the use of chemicals or materials that are permanently applied over the metal. These finishes are divided into mill, or as fabricated, finishes and processed mechanical finishes.

Chemical finishes include decorative finishes, but most so-called chemical finishes are actually cleaning procedures or treatments carried out to prepare a metal to receive other finishes. As the name implies, chemical finishes are produced by bringing a metal into contact with chemicals that change it chemically or physically.

Coatings are applied finishes, which fall into two categories: inorganic and organic. The inorganic coatings include metallic, anodic, vitreous, and laminated coatings. Organic coatings include both organic and resinous materials such as paint, lacquer, and enamel.

This chapter addresses mechanical and chemical finishes, and metallic, anodic, vitreous, and laminated coatings that might be used on nonstructural metal items. It includes discussions of the materials, their application, possible failures that may occur in them, and methods the industry recommends for repairing those finishes.

Organic coatings are discussed in Chapter 5.

Chapter 3 discusses the metals that the finishes in this chapter may be applied over.

The terms used in this chapter are used as defined in the Glossary.

Mechanical Finishes for and Cleaning of Ferrous Metals

Mechanical finishes include those left by the manufacturing process (mill finishes) and those created by cold rolling with polished rollers and by polishing and buffing. Standards for mechanical finishes on iron and steel have been established by the American Iron and Steel Institute (AISI). They are addressed at some length in the National Association of Architectural Metal Manufacturers' *Metal Finishes Manual for Architectural and Metal Products*.

Mechanical cleaning includes that used to remove mill scale and oils and to prepare metal to receive applied finishes.

Mill Finish

A mill finish is simply the natural finish that the metal has as it leaves the initial manufacturing process.

Mill Finish on Iron and Carbon Steel. Except in rare instances, mill-finished iron and carbon steel is not suitable for use without further treatment. The mill finish produced by hot rolling, which is called a black, as-rolled finish, carries a coating of scale and rust. Steel produced by the cold-rolled process is almost always coated with grease and oil and is often so smooth that it will not hold paint well.

Mill finishes on iron and steel are not generally permitted to stand without further finishing. The processes discussed in "Preparation of Metal Surfaces" under the heading "Organic Coating Application" in Chapter 5 are the usual mechanical finishes for iron or steel, but they are properly called cleaning rather than finishing.

The operating parts of machines may be polished and used without an applied finish when they will be protected in use by coats of oil or grease. Unprotected ordinary iron and carbon steel will soon corrode and become useless, but some types of iron and steel are less likely to corrode, and some corrode but their corrosion forms a protective covering. Real wrought iron, as opposed to the mild steel used today, resists rusting more than most iron or steel products. Some modern low-alloy copper-bearing steels are recommended for use exposed, without a coating. They oxidize, but the dark brown compounds that form are dense and, unlike common rust, adhere tightly to the steel, forming a protective shell that resists further oxidation.

Mill Finish on Stainless Steel. The mill finish on stainless steel sheet and strip varies, depending on whether the steel was hot or cold rolled. A hot-rolled mill finish, which is comparatively rough and dull, is called a No. 1 Sheet Finish. It is produced by the rolling process, followed by annealing and descaling.

There are two levels of sheet or strip finishes that can be achieved by the cold-rolling process. The first, called No. 2D Sheet Finish or No. 1 Strip Finish, is produced by rolling, followed by descaling, pickling, and a final pass through unpolished rollers.

The second, which is called No. 2B Sheet Finish or No. 2 Strip Finish, occurs when stainless steel with a No. 2D Sheet Finish or No. 1 Strip Finish is passed through highly polished rollers. It is more often used in building and related construction than is the No. 2B Sheet or No. 2 Strip Finish.

A finish called Bright Annealed Finish can be obtained by annealing stainless steel sheet and strip that has been given a No. 2B Sheet or No. 2 Strip Finish.

The standard mill finish for stainless steel plates is dull and nonreflective. Bars have a special mill finish that is applicable only to them. Pipe and tubing may have a mill finish that resembles No. 1 Strip Finish if they are hot rolled or forged or the No. 2 Strip Finish if they are then further finished by cold rolling. Extrusions have a mill finish that resembles No. 1 Sheet Finish.

Polished, Patterned, and Other Stainless Steel Finishes

Stainless steel is often polished mechanically as a part of the finishing operation. Sometimes a patterned finish is applied. Nonstandard finishes are also available.

Polished Stainless Steel. Five levels of polished finish are commonly used on stainless steel used in building and related construction. These are

No. 3, which is a semifinished surface that is usually polished more highly in a final product; No. 4, a general-purpose bright polished finish that is probably the most common stainless steel finish used in building and related construction; No. 6, which is a widely used soft satin finish; No. 7, a reflective satin finish; and No. 8, a mirrorlike finish.

These five finish designations apply primarily to sheet materials, but the same polished finishes are also available on plate, strip, bars, pipe and tubing, and extrusions.

Patterned Stainless Steel.   A pattern may be imparted to stainless steel sheets by passing them between matched rollers that have the desired design embossed on them, or by a variation of this process. Many patterns are available.

Other Mechanical Finishes for Stainless Steel.   Various producers provide stainless steel sheets with patterns imposed on them. Cross-brushed, matte, frosted, geometrically patterned, and many other variations are available. The production methods include various types of cold rolling, polishing, and grinding.

Mechanical Cleaning of Ferrous Metals

There are two basic classes of mechanical cleaning methods used on ferrous metals other than stainless steel. The methods in the first class will effectively remove mill scale and rust, but not grease and oil. This class includes the following two methods recommended by the Steel Structures Painting Council (SSPC):

Hand-tool cleaning in accordance with SSPC-SP-2. As its name implies, this method is carried out using hand tools. It is highly labor intensive and is therefore best suited for use in spot cleaning.

Power-tool cleaning in accordance with SSPC-SP-3. The grinders, sanders, brushes, and abrasives used in this method will often so damage thin metals that they will become useless. Its use is therefore generally limited to thick materials.

The methods in the second class, sand and shot blasting, are probably the best techniques for removing mill scale and rust. They will also remove oil and grease and roughen the surface, giving it enough "tooth" for paint to adhere well. This class includes the following four methods recommended by the SSPC:

White Metal Blast Cleaning in accordance with SSPC-SP-5.

Commercial Blast Cleaning in accordance with SSPC-SP-6.
Brush-Off Blast Cleaning in accordance with SSPC-SP-7.
Near-White Blast Cleaning in accordance with SSPC-SP-10.

Two cleaning methods formerly recommended by the SSPC, flame cleaning (SSPC-SP-4) and weathering (SSPC-SP-9), are now obsolete and are no longer recommended by the SSPC.

Chemical Finishes for and Cleaning of Ferrous Metals

Chemicals are mostly used to prepare metals for application of other finishes, but some are used decoratively, as discussed in the following sections.

Chemical Cleaning and Pretreatment of Steel

Chemical Cleaning of Carbon Steel.   There are essentially four chemical-cleaning methods in general use: solvent cleaning (SSPC-SP-1), pickling (SSPC-SP-8), vapor degreasing, and alkaline degreasing.

Pickling removes mill scale and rust by immersing the metal in a dilute acid solution. Vapor degreasing consists of exposing the metal to chemical vapors. Alkaline degreasing involves immersing the metal in, or spraying it with, an alkaline solution. Sometimes mechanical brushing is also used.

Chemical Cleaning of Stainless Steel.   Bare stainless steel that is to receive an organic finish should be washed with solvent and prepared according to the paint or coating material manufacturer's recommendations.

Pretreatment of Carbon Steel.   Carbon steel is usually given a conversion coating to change (convert) the chemical nature of the surface so that paint and coatings will adhere to it more readily. Acid phosphate solutions are the materials most commonly used to produce these conversion coatings.

Chemical Decoration of Stainless Steel

Conversion coatings are used to darken, also called blacken, stainless steel. The resulting color may be blue, dark brown, or black, depending on the coating and process used. The conversion coating causes an oxide to form on the stainless steel, which is the source of the color.

Another coloring effect can also be obtained by flash coating stainless steel with nickel or copper.

Requirements Applicable to All Inorganic Nonferrous Metal Finishes

All finishes on aluminum and copper alloys, except for the category of as fabricated finishes, are called "process finishes," with one possible exception. Aluminum sheets that have been given an organic or laminated coating in the mill (coil coated) fall into a gray area. They are called process finishes by some people and mill finishes by others.

Standards for aluminum finishes have been established by the Aluminum Association (AA) and for copper alloys by the National Association of Architectural Metal Manufacturers (NAAMM). Since no purpose would be served by repeating the designation systems here, this discussion is a summary only. The standards for finishes on aluminum and copper alloys are addressed fully in the NAAMM publication *Metal Finishes Manual for Architectural and Metal Products.* Anyone who must deal with new or existing nonstructural metals or their finishes would do well to obtain a copy of it. Determining the specific category into which an existing aluminum or copper alloy finish falls may be necessary if matching is required. Identifying specific finishes within the types included in the standard designations is, however, a job for professionals.

Finish Designations for Aluminum

The Aluminum Association's designated types of mechanical finishes for use on aluminum include as fabricated, buffed, directional textured, non-directional textured, and patterned finishes. Within these types there are many finishes. Each finish in the first four categories is given a designation that consists of the letter M followed by a two-digit number. The first digit corresponds to the four types of finishes. The second digit designates the specific finish within that type. Thus, M21 is a smooth, specular, buffed finish and M31 is a fine satin, directional (scratch lines all in the same direction) textured finish.

The chemical finishes for use on aluminum are similarly designated. The four types are non-etched cleaned, etched, brightened, and conversion coatings. The letter used with this system is C. The first digit corresponds to the four types of finishes. The second digit designates the specific finish. Thus, C21 is a fine matte, etched finish and C31 is a highly specular, brightened finish.

Anodic coatings are similarly designated, using the letter A. The first digit here corresponds to the headings general, protective and decorative, architectural class II, and architectural class I. Thus, A31 is a clear architectural class II anodic coating, A41 a clear architectural class I anodic coating.

Other coatings are similarly denoted. The letters are R for resinous and other organic coatings, V for vitreous coatings, E for electroplated and metallic coatings, and L for laminated coatings.

When an x appears in a designation, it is necessary to follow that designation with a description, usually of a proprietary finish.

All the designations are used together when specifying a particular finish. The entire designation is preceded by AA- to identify it as an Aluminum Association designation. For example, the finish AA-M22C22A42 has a specular, buffed, mechanical finish (M22), a medium matte, etched, chemical finish (C22), and an integrally colored, architectural class I anodic coating (A42).

Mechanical Finishes for Aluminum

Mechanical finishes include those left by the manufacturing process and ones created by grinding, polishing, sand blasting, and rolling.

As Fabricated Finish for Aluminum

An as fabricated (mill) finish is simply the natural finish that is imparted by the casting, rolling, or extruding process. It is the finish the aluminum has when it leaves the initial manufacturing process. It is not the final finish left after the material has been fabricated into a nonstructural metal item of the sort discussed in Chapters 6 and 7. The term "as fabricated finish" used with aluminum corresponds to the term "mill finish" used for iron and steel. When speaking of aluminum finishes the terms "mill" and "as fabricated" are used interchangeably.

Most aluminum with an as fabricated finish will display some imperfections, which must be taken into account when deciding whether to use it without further finishing. Castings, for example, have a rough, matte finish, sand castings being rougher than die castings. Hot-rolled aluminum may be darker than cast, cold-rolled, or extruded aluminum and will usually have some random discolorations as well. Extrusions are usually the same color as cold-rolled material but are often left with parallel striations called die lines. Most imperfections tend to become worse in appearance when the metal is formed further and assembled into nonstructural metal items of the kinds addressed in this book.

Cold-rolled aluminum items usually come closer to the final appearance desired than do those produced by other methods. This process permits some control by varying the amount of polishing used on the rollers. The smoother the rollers are, the closer a cold-rolled surface will be to a desirable finish. Even cold-rolled aluminum, however, may have stains, or a coat of oil, or both.

The Aluminum Association divides as fabricated finishes on aluminum into four classes:

Unspecified (M10).
Specular as fabricated (M11).
Non-specular as fabricated (M12).
Other (M1x).

An unspecified as fabricated finish is the natural finish imparted by casting, hot rolling, cold rolling with unpolished rollers, or extruding.

A specular as fabricated finish is a mirrorlike finish produced by cold rolling with polished rollers. The specular finish may be applied to one or both sides. Castings, forgings, and extrusions cannot be given a specular finish.

A non-specular as fabricated finish is more uniform than an unspecified finish, but it lacks the mirrorlike quality of a specular as fabricated finish.

The Other category is reserved for special finishes that do not fit into one of the other three categories.

Other Mechanical Finishes for and Cleaning of Aluminum

Buffed Finishes. Aluminum finishes created by a process of buffing alone—or by grinding, polishing, and buffing—are called buffed finishes. The Aluminum Association lists them in four categories:

Unspecified (M20).
Smooth specular (M21).
Specular (M22).
Other (M2x).

An unspecified buffed finish is optional with the aluminum finisher.

A specular buffed finish is produced by buffing alone. A smooth specular buffed finish, the brightest and most lustrous mechanical finish that can be developed on aluminum, is produced by successively grinding, polishing, and buffing.

The Other category is reserved for special finishes that do not fit into one of the other three categories.

Directional Textured Finishes for Aluminum. The Aluminum Association designation system recognizes seven standard finish types in the Directional Textured Finish category. They are:

Unspecified (M30).

Fine satin (M31).

Medium satin (M32).

Coarse satin (M33).

Hand rubbed (M34).

Brushed (M35).

Other (M3x).

An unspecified directional textured finish is optional with the aluminum finisher.

The three satin finishes represent different degrees of fineness. They are all produced by wheel or belt polishing. In spin finishing an abrasive cloth is held against the metal and rotated, followed by stainless steel wool or emery polishing. This produces a bright finish with a fine pattern of concentric circles.

The hand rubbed category is an expensive type of finish produced by rubbing the metal with abrasive cloths or stainless steel wool. It is seldom used, except for final touch-ups on other satin finishes.

The brushed category is a directional finish characterized by fine parallel scratches made by stainless steel wire brushes, sander heads, impregnated plastic disks, and other means.

The Other category is reserved for special finishes that do not fit into one of the other six categories.

Non-Directional Textured Finishes for Aluminum.   Finishes that the Aluminum Association designated as non-directional textured ones are produced by abrasive blasting. This category includes nine subdivisions as follows:

Unspecified (M40).

Extra fine matte (M41).

Fine matte (M42).

Medium matte (M43).

Coarse matte (M44).

Fine shot blast (M45).

Medium shot blast (M46).

Coarse shot blast (M47).

Other (M4x).

Most of these finishes are matte in appearance, with various degrees of roughness. They are not widely used on the types of metal items addressed in this book. Their most extensive use is on castings. They should not be used on metal that is 1/4 inch or less thick, because they tend to distort the metal.

An unspecified non-directional textured finish is optional with the aluminum finisher.

The categories with "matte" in their names are produced by blasting with sand or aluminum oxide. Those with "shot blast" in their names are produced by steel shot blasting.

The Other category is reserved for special finishes that do not fit into one of the other categories.

Combinations are sometimes used on surfaces that have relief in which incised areas are abrasive blasted while high areas are masked. Then the high areas are polished.

Patterned Finishes for Aluminum. Thin aluminum sheets can be given a patterned finish by rolling as fabricated sheets between rollers shaped to the desired design. There is no specific Aluminum Association category for patterned finishes.

Mechanical Cleaning of Aluminum. After being washed with mineral spirits or turpentine, bare new aluminum that is to receive an organic finish should either be allowed to weather for one month or be roughened with stainless steel wool.

Chemical Finishes for and Cleaning of Aluminum

Chemicals are used to prepare aluminum for the application of other finishes, to act themselves as a final finish, or to be part of a total finishing process involving other steps.

The Aluminum Association lists four types of chemical finishes for aluminum: non-etched cleaned, etched, brightened, and chemical conversion coatings.

Non-Etched Cleaned Aluminum

There are a number of chemical methods commonly used to clean aluminum without otherwise altering the metal. Some such cleaning is essential if the aluminum is to receive an applied finish. The Aluminum Association lists four categories, of which the middle two are the most commonly used. The four are:

Unspecified (C10).

Degreasing (C11).

Chemical cleaning (C12).

Other (C1x).

An unspecified non-etched cleaned finish is optional with the aluminum finisher.

Degreasing, which could be called vapor degreasing, is exposing the aluminum to the vapors of a chlorinated solvent in a machine designed for that purpose.

Chemical cleaning is sometimes used alone but more commonly is a second-stage treatment following degreasing. This process is sometimes called "inhibited" chemical cleaning, because the chemicals used are inhibited, to prevent them from etching the aluminum's surface.

The Other category is reserved for special finishes that do not fit into one of the other categories.

Bare aluminum, especially existing aluminum, that is to be field painted before being subjected to weathering or stainless steel wool roughening, as indicated earlier in this chapter, should be washed with mineral spirits or turpentine.

Etched Aluminum

The Aluminum Association divides etched finishes into five categories:

Unspecified (C20).

Fine matte (C21).

Medium matte (C22).

Coarse matte (C23).

Other (C2x).

In essence, an etched finish is a matte finish produced by dipping, washing, or otherwise exposing aluminum to an alkali or acid solution. The different degrees of matte finish are accomplished by using different chemicals.

An unspecified etched finish is optional with the finisher.

The Other category is reserved for special finishes that do not fit into one of the other categories.

Etching is a very common process in aluminum production. The popular finish called "frosted" is made in this way by using caustic soda (sodium hydroxide) etching. Etching is often used as a pretreatment before anodizing. Sometimes, etched aluminum is simply given a clear lacquer coating and used without further finishing.

Brightened Aluminum

The Aluminum Association classifies brightened finishes into four categories:

Unspecified (C30).
Highly specular (C31).
Diffuse bright (C32).
Other (C3x).

There are two methods of brightening aluminum: chemical brightening and electro-brightening. Both produce surfaces that are mirror bright. These finishes are seldom used in nonstructural metal items of the types discussed in this book.

An unspecified brightened finish is optional with the finisher.

The Other category is reserved for special finishes that do not fit into one of the other categories.

Conversion Coatings for Aluminum

A natural oxide forms a thin film on the surface of bare aluminum, which prevents applied finishes from bonding. The application of certain chemicals changes (converts) the chemical nature of the oxide film so that it provides a good bond for paint, organic coatings, and laminates. The products of such chemical treatment are called conversion coatings or conversion films.

The Aluminum Association classifies conversion coatings into the following five categories:

Unspecified (C40).
Acid chromate-fluoride (C41).
Acid chromate-fluoride-phosphate (C42).
Alkaline chromate (C43).
Other (C4x).

Conversion coatings on aluminum are produced by proprietary chemicals. Those used to produce C41-type conversion coatings may leave a clear or green color, depending on the actual chemical used. The C42 coatings may also be clear, but they are sometimes yellowish. The C43 coatings, which are gray, are sometimes allowed to stand as the aluminum's final finish.

An unspecified conversion coating is optional with the aluminum finisher.

The Other category is reserved for conversion coatings that do not fit into one of the other categories.

An excellent mechanical paint bond can be achieved on aluminum simply by etching the surface with a phosphoric acid solution.

Finish Designations for Copper Alloys

The NAAMM's designated types of mechanical finishes for use on copper alloys include as fabricated, buffed, directional textured, non-directional textured, and patterned. Within these types there are many finishes. Each finish in the first four categories is given a designation that consists of the letter M followed by a two-digit number. The first digit corresponds to one of the five types of finish, and the second digit designates the specific finish within that type. Thus, M21 is a smooth specular, buffed finish, M31 a fine satin, directional textured finish.

The chemical finishes for use on copper alloys are similarly designated, but there are only two types: non-etched cleaned and conversion coatings. The code letter used is C. The first digit corresponds to the finish types, the second to the specific finish. Thus, C11 is a degreased cleaning, and C51 is a cuprous chloride–hydrochloride acid patina.

Coatings are similarly denoted. The letter O is used for clear organic coatings, L for laminated coatings.

When an x appears in a designation, it is necessary to follow that designation with a description of the finish, usually of a proprietary one.

All the appropriate designations are used together when describing a particular finish. For example, the finish M31-M34-07x has a fine satin (M31), hand rubbed (M34), directional textured mechanical finish and a clear, thermoset organic coating (07x). To make the designation complete, the organic coating, represented by the x, must be described completely, including its characteristics and manufacturer.

Mechanical Finishes for Copper Alloys

Mechanical finishes include those left by the manufacturing process and those created by grinding, polishing, sand blasting, and rolling.

As Fabricated Finish for Copper Alloys

An as fabricated finish is simply the natural finish that is imparted by the casting, rolling, or extruding process. It is the finish a copper alloy has when it leaves the initial manufacturing process, not the final finish after the material has been fabricated into a nonstructural metal item of the sort discussed in Chapters 6 and 7. The term "as fabricated finish" corresponds to the term "mill finish" used for iron and steel.

Most copper alloys with an as fabricated finish display some imperfections that must be taken into account when deciding whether to use them without further finishing. Castings, for example, have a rough matte finish, with sand castings being rougher than die castings. Hot-rolled copper-alloy items will have a dull surface, be darker than cast, cold rolled, or extruded copper-alloy items, and will usually have some random discolorations. Extrusions are usually the same color as cold-rolled material, but they are often left with parallel striations called die lines. Most imperfections will tend to become worse in appearance when the metal is formed further and assembled into nonstructural metal items of the kinds addressed in this book.

Cold-rolled copper-alloy items usually come closer to the final appearance desired than those produced by other methods. The process permits some control by varying the amount of polishing used on the rollers—the smoother the rollers, the closer the cold-rolled surface will be to a desirable finish. Even cold-rolled copper-alloy items, however, may have stains, a coat of oil, or both.

The NAAMM divides as fabricated finishes on copper alloys into four classes:

Unspecified (M10).

Specular as fabricated (M11).

Matte finish as fabricated (M12).

Other (M1x).

An unspecified as fabricated finish is the natural finish imparted by casting, hot rolling, cold rolling with unpolished rollers, or extruding.

A specular as fabricated finish is a mirrorlike finish produced by cold rolling with polished rollers. The specular finish may be applied to one or both sides. Castings, forgings, and extrusions cannot be given a specular finish.

A matte as fabricated finish is a dull finish produced by annealing copper alloy items that have been produced by hot rolling, cold rolling with unpolished rolls, casting, or extruding.

The Other category is reserved for special finishes that do not fit into one of the other categories.

Other Mechanical Finishes for and Cleaning of Copper-Alloy Items

Buffed Finishes.   Copper-alloy finishes created by a process of grinding, polishing, and buffing are called buffed finishes. There are four standard buffed finishes for copper alloys:

Unspecified (M20).
Smooth specular (M21).
Specular (M22).
Other (M2x).

An unspecified buffed finish is optional with the copper alloy finisher. A smooth specular, buffed finish, which is the brightest and most lustrous mechanical finish that can be developed on copper-alloy items, is produced by successive grinding, polishing, and buffing.

A specular buffed finish is produced by the same methods as a smooth specular, buffed finish but is not as smooth.

The Other category is reserved for special finishes that do not fit into one of the other categories.

Directional Textured Finishes for Copper Alloys. The standard designation system recognizes eight finish types in the directional textured finish category, which are:

Unspecified (M30).
Fine satin (M31).
Medium satin (M32).
Coarse satin (M33).
Hand rubbed (M34).
Brushed (M35).
Uniform (M36).
Other (M3x).

An unspecified directional textured finish is optional with the copper alloy's finisher.

The three satin finishes represent different degrees of fineness. They are all produced by wheel or belt polishing. Spin finishing produces a bright finish with a fine pattern of concentric circles. In this process an abrasive cloth is held against the metal and rotated, followed by stainless steel wool or emery polishing.

The hand rubbed category is an expensive finish produced by rubbing the metal with a fine brass brush or nonwoven abrasive mesh pads, abrasive cloths, or stainless steel wool. It is seldom used for general finishing but often for final touch-up on the other satin finishes.

The brushed category is a directional finish characterized by fine parallel scratches produced by stainless steel, nickel steel, or brass wire brushes, sander heads, impregnated plastic disks, and other means.

The uniform category finish is produced by a single pass of a number 80 grit belt.

The Other category is reserved for special finishes that do not fit into one of the other categories.

Non-Directional Textured Finishes for Copper-Alloy Items. Finishes designated as non-directional textured are produced by abrasive blasting. This category includes nine finishes, as follows:

Unspecified (M40).

Unassigned (M41).

Fine matte (M42).

Medium matte (M43).

Coarse matte (M44).

Fine shot blast (M45).

Medium shot blast (M46).

Coarse shot blast (M47).

Other (M4x).

Most of these finishes are matte in appearance, with various degrees of roughness. They are not widely used on the types of metal items covered in this book. Their most extensive use is on castings. They should not be used on metal that is 1/4 inch or less in thickness, because they tend to distort the metal.

An unspecified non-directional textured finish is optional with the finisher.

The categories with "matte" in their names are produced by blasting with silica sand or aluminum oxide. Those described as "shot blast" are produced by steel-shot blasting.

The Other category is reserved for special finishes that do not fit into one of the other categories.

Combinations are sometimes used on surfaces that have relief in which the incised areas are abrasive blasted while the high areas are masked. Then the high areas are polished.

Patterned Finishes for Copper-Alloy Items. Thin copper-alloy sheets can be given a patterned finish by rolling as fabricated sheet between rollers shaped to the desired design. There is no specific standard category for patterned finishes.

Mechanical Cleaning of Copper Alloys. After bare copper and copper-alloy material that is to receive an organic finish has been washed with mineral spirits, it should have stains, mill scale, and other foreign material removed by sanding.

Chemical Finishes for and Cleaning of Copper Alloys

There are two primary reasons that chemicals are used on copper alloys; to clean them of foreign matter, and to change the metal's color and provide a final finish. The two chemical finishes we are concerned with are called non-etched cleaned and conversion coatings. There are other chemical treatments used on copper alloys, but they are not extensively used in the kinds of items we are interested in here.

Non-Etched Cleaned Copper Alloys

There are a number of chemical methods commonly used to clean copper alloys without otherwise altering the metal. Some such cleaning is required if the metal is to receive another finish. The four standard categories of such methods are:

Unspecified (C10).

Degreased (C11).

Cleaned (C12).

Other (C1x).

An unspecified non-etched cleaned finish is optional with the finisher.

Vapor degreasing consists of exposing the metal to the vapors of a chlorinated solvent in a machine designed for that purpose.

Chemical cleaning is sometimes used alone, but it is more commonly a second-stage treatment following degreasing. This process is sometimes called "inhibited" chemical cleaning, because the chemicals used are inhibited to prevent them from etching the metal's surface.

The Other category is reserved for special finishes that do not fit into one of the other categories.

Bare copper and copper-alloy items, especially existing ones, that are to be field painted before being sanded, as indicated earlier in this chapter, should be washed with mineral spirits to remove dirt, grease, and oil.

Conversion Coatings for Copper Alloys

Conversion coatings are used to change the color of a copper alloy and provide a final finish for the metal. As a copper alloy ages it oxidizes, changing its color and producing a coating that not only modifies the appearance of the material but protects it from further oxidization. Conversion coatings are a not always completely successful attempt to duplicate a normal aged appearance and form a protective coating by accelerated chemical means. The coatings produced are oxides or sulfides of the metal.

There are two basic types of conversion coatings commonly used: those that produce a patina (verde antique) finish and those that produce an oxidized finish known as statuary bronze.

There are seven types of materials used to produce the more common conversion coatings. The following list includes those types, the NAAMM designation for each, and the type of conversion coating that each chemical produces. The list also includes an Other category for materials not listed.

Ammonium chloride (C50) patina.

Cuprous chloride–hydrochloric acid (C51) patina.

Ammonium sulfate (C52) patina.

Carbonate (C53) patina.

Oxide (C54) statuary bronze.

Sulfide (C55) statuary bronze.

Selenide (C56) statuary bronze.

Other (C5x).

Patinas, which are produced using acid chlorides, are more difficult to control than statuary bronze finishes. Patinas often have variations in color, especially over large surface areas, and sometimes fail to adhere to the metal. They are also likely to stain adjacent materials.

Oxidized (statuary) finishes are somewhat more stable than patinas, and their color is easier to control. They come in three tones: light, medium, and dark. A range of color should be expected, however, even within a single tone group.

The Other category is reserved for conversion coatings that do not fit into one of the other categories.

Other Chemical Finishes for Copper Alloys. There are several other chemical finishes given to copper alloys, but they are not used extensively for the types of nonstructural metal items used in building and related construction. Two of them are those known as bright finish and matte dipped finishes. Both are created by dipping the metal in chemicals.

The bright finish is produced by an acid dip. The matte dip finish is a secondary dipping process. These processes produce a proper base for other finishing, such as plating and organic coatings.

Acid etching is also used on copper alloys—but as a means of producing surface designs or patterns, not as a true finishing method.

Metallic Finishes

Some sheet metals used in building and related construction are plated or coated with another metal. The base metal may be either ferrous or nonferrous, but the coating or plating metal is usually nonferrous. In most cases the two metals are bonded together, but they remain separate metals and do not form an alloy. In an alloy the constituent materials are melted together to form a single material. For example, the various types of steel are all alloys of iron and other materials. Bronze is an alloy of copper.

Finishes are applied to metals for two purposes: to protect the metal and to decorate it. Some metallic finishes, such as galvanizing and aluminum plating on steel, are primarily intended only to protect the metal and do not serve to decorate it, although they may be left as the final finish in inconspicuous locations. When decoration is desired, these kinds of metallic finishes are usually given a finish coat of another material. Other metallic finishes such as chromium plating both decorate and protect the underlying metal.

Metal-coated steel is characterized by having the strength of the steel and the corrosion resistance of the metal used in the coating, but the different metals used to coat steel do not all protect it in the same way. Zinc and cadmium, for example, form electrochemical barriers that protect steel from corrosion by themselves corroding (Fig. 4-1). Aluminum, copper, nickel, chromium, terne, tin, and other similar metal coatings protect steel by forming a barrier that prevents oxygen and other corrosive materials, such as acid rain and other chemicals from reaching the steel. When these

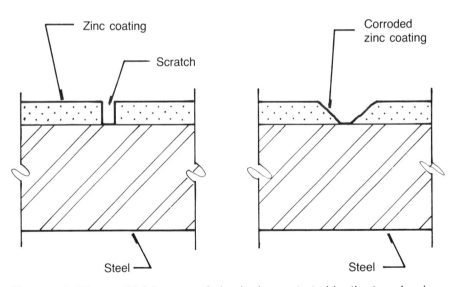

Figure 4-1 The sacrificial nature of zinc is demonstrated by the two drawings in this figure. A small scratch occurs in the galvanizing. The zinc coating then corrodes (sacrifices itself), thus preventing the underlying steel from beginning to rust until the corrosion in the zinc is quite far advanced.

types of metal coatings are damaged, however, they do not perform the kind of sacrificial action that characterizes zinc. Instead the steel will rust beneath its coating (Fig. 4-2). The effect is similar to that of rust beneath an organic coating, as shown in Figure 4-3, except that the swelling and bulging of the finish shown in that photograph will probably not occur in a metal coating. The ongoing damage may therefore be more difficult to detect.

Metal coatings on nonferrous metals are used mostly to give the underlying metal the desired finish, although some also protect it. Chromium is much harder and more scratch resistant, for example, than either aluminum or copper, which it is often used to coat. Some metallic coatings protect other materials from the coated metals. Lead-coated copper, for instance, prevents water runoff from the copper from staining, corroding, or otherwise damaging adjacent materials.

There are six basic methods of coating one metal with others: hot dipping, electroplating, spraying (metallizing), cladding, alloying (cementation), and fusion welding.

In the hot-dip process, the underlying metal is coated with a second metal by being immersed in a molten bath of the second metal. Zinc, aluminum, tin, and terne are thus applied to steel, for example.

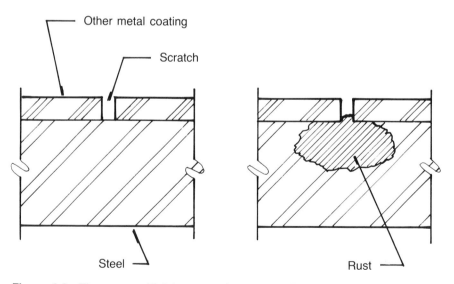

Figure 4-2 The nonsacrificial nature of most metal coatings is demonstrated by the drawings in this figure. The same scratch has occurred here as in Figure 4-1, but in this case the coating does not sacrifice itself, thus permitting the steel to rust beneath the coating.

Figure 4-3 An example of corrosion occurring beneath a coating.

Zinc, cadmium, aluminum, and nickel are often deposited on steel, aluminum, and copper alloy by the electroplating process. Chromium and copper are similarly deposited over nickel that has itself been deposited on steel or aluminum. Chromium is also deposited over nickel deposited over copper over steel.

Most metal coatings can be applied to steel and nonferrous metals by spraying, but probably the most frequent use of this method is to apply zinc and aluminum to steel. This is the only method, other than touchup galvanizing repair by brush, that may be effectively used in field applications.

Steel is covered with copper, stainless steel, and aluminum by the cladding process. This technique, which may include dipping, rolling, electrolytic depositing, or electrowelding, differs depending on the metal that forms the cladding, but in all cases the cladding metal and the steel remain as separate materials.

Cementation is the forming of an alloy between a coating metal and the coated metal. When the coated metal is steel, this process may be called chromizing, carronizing, Sherardizing, calorizing, or Ihrigizing, depending on the coating material and method used.

Fusion welding is, as it sounds, the fusing of the cladding metal to the base metal by welding.

The most common methods used in the building industry to coat steel with other metals are the hot-dip and electroplating processes. Galvanizing (zinc coating) and aluminizing (aluminum coating) are the most common coatings used on steel.

The most common method used to coat nonferrous metals with metal for use in the building industry is electroplating, although some dipping is also used.

Galvanized Steel

Material. Zinc is a bluish-white crystalline metal. Steel that has been coated with zinc, either electrolytically or by the hot-dip process, is called galvanized steel. Most galvanized steel used in building and related construction, especially on materials and products that will be painted in the field or left exposed without paint, is produced by the hot-dip process. Some shop- or factory-fabricated and finished nonstructural metal items, however, are made using metal on which zinc has been deposited by electroplating.

The general requirements for hot-dip galvanized steel-sheet products are contained in ASTM Standard A 525. Hot-dip galvanized steel sheet is available in several qualities, including commercial, lock forming, drawing, special killed (deoxidized, and thus holefree) drawing, and structural (physical quality) sheet.

Several types of zinc coatings are available, including regular spangle, minimized spangle, iron–zinc alloy, wiped, and differential. The most commonly used type of zinc coating for nonstructural metal items is regular spangle, but other types may also be used. Different zinc-coating designations are used for each type of coating. Coating designations for the regular spangle type begin with a G prefix, for example. The most commonly used coating designation for sheet material used to produce the types of nonstructural metal items addressed in this book is G 90, which requires that the coating have 0.90 ounces of zinc on each square foot of metal surface. Designation G 60, which requires 0.60 ounces per square foot, is sometimes used where the metal is to be in a more protected location.

Materials with other coating thicknesses (designations) are also used occasionally. For extreme environments, zinc coatings of up to 2.35 ounces per square foot may be used. In all cases, the zinc-coating weights for sheet materials refer to the total weight of the coatings on both sides of the sheet combined.

Most galvanized sheet steel used in the construction industry has a regular spangle finish, but some is treated further by being wipe coated or galvannealed. Wipe-coated sheets are wiped down as they are withdrawn from the molten zinc, which removes the spangle and leaves a thin zinc–iron alloy coating. Galvannealing is a process of heat treating that when used on hot-dip galvanized sheets removes spangle and leaves a gray, zinc–iron alloy surface.

Coating weights for galvanized steel products other than sheets are specified in the ASTM standard applicable to the galvanizing of each product. They vary from product to product. For example, a common zinc weight for plates and shapes 1/4 inch or less in thickness is 2 ounces per square foot, for thicker plates and shapes 2.3 ounces per square foot, for thick hardware items 2 ounces per foot, and for thin hardware 1.5 ounces per foot. Many other thicknesses may also be used, depending on the item being galvanized and the degree of hazard involved.

The galvanized steel products most commonly used to fabricate the types of nonstructural metal items addressed in this book are shown in the following list. The standard listed following each item is the usual standard for that product.

Pipe: ASTM Standard A 53.

Rolled, pressed, or forged steel shapes, plates, bars, and strip 1/8 inch thick and heavier: ASTM Standard A 123.

Iron and steel hardware: ASTM Standard A 153. This category includes castings; rolled, pressed, and forged articles; bolts and their nuts and washers; and screws, rivets, nails, and similar items.

Assembled steel products: ASTM Standard A 386.

Structural steel sheet: ASTM Standard A 446.

Commercial quality carbon steel sheet: ASTM Standard A 526.

Carbon steel wire: ASTM Standard A 641.

Other galvanized steel products may also be used. The ASTM standards for other types of galvanized steel products are listed in the Bibliography and marked with a [4].

Cleaning and Chemical Pretreatment and Decorating of Galvanized Steel. Unprotected galvanized metal will form a coating called "white rust," which can be removed by wire brushing, sanding, or blasting. If normal rust appears on a galvanized surface, this indicates a failure of the galvanizing. In this case the rust should be removed, exposing the bare metal. The galvanizing should then be repaired using one of the available galvanizing repair paints that are specifically formulated for the purpose.

To help prevent white rust, fabricators often coat galvanized surfaces with oils, waxes, silicons, or silicates. Everyone agrees that such coatings must be removed before an organic finish can be applied. They disagree, however, about how to do so. Some sources say that bare galvanized surfaces should be cleaned free of oil and other surface contaminants by using mineral spirits or xylol. Others recommend using a solvent wash and specifically recommend against using mineral spirits. Some say to use petroleum spirits. Others say not to use petroleum-based solvents.

Galvanized steel is often treated with a process called "bonderizing," which leaves the surface ready for immediate painting. Bonderizing consists of a hot-phosphate treatment that leaves a crystalline zinc-phosphate film on the surface.

Authorities disagree about the proper pretreatment of galvanized metal to receive paint. For example, some sources say that it should be permitted to weather for at least six months before it is painted. Other experts say that weathering is a bad idea, because galvanized metal weathers unevenly, which can cause poor paint adhesion. They argue that unless the metal is directly exposed it will not weather appreciably anyway, that even more preparation is required after the metal has been allowed to weather, and that weathering will make the preparation more difficult to accomplish properly.

Most sources say that after all oils, white rust, and other contaminants have been removed, bare galvanized metal should be pretreated before the first coat of an organic finish is applied. Galvanized steel that is to be factory coated is often pretreated using complex oxides, zinc phosphates, or chromates in accordance with ASTM D 2092. However, even the industry sources that recommend pretreatment before field painting disagree on the proper material to use for it. Some say to use a weak acetic acid, others a proprietary acid-bound resinous or crystalline zinc-phosphate preparation or phosphoric acid. Still other sources recommend that acetic acid not be used on galvanized metal. The only option left when trying to decide which pretreatment to use—or even whether to use a pretreatment at all—is to ask for a recommendation from the manufacturer of the paint that will be used. At least then there will be someone to complain to if the paint fails.

Some conversion coatings are used to dye zinc-plated steel, which is particularly effective when the item is articulated. In this process the dyed material is polished so that its raised portions are bright metal (zinc), while the depressions remain dark, in the color of the dye.

Aluminized Steel

Aluminum is a light, bluish, silver-white metal that is ductile, malleable, and highly resistant to oxidation. Aluminized steel consists of a steel sheet that has been coated with aluminum. This product is generally expected to conform with the requirements of either ASTM standards A 792, A 875, or A 463. The latter is the one most often used in prepainted applications. To conform with Standard A 792 the coating must be 55 percent aluminum and 45 percent zinc. A coating conforming with Standard A 875 would be 95 percent zinc and only 5 percent aluminum and misch metal. Standard A 463 requires a minimum coating weight of aluminum of 0.65 ounces per square foot. Most aluminized steel sheet is between 0.12 and 0.046 inches thick, but both thicker and thinner materials may be used.

Aluminized steel is a sheet product used primarily in preformed and formed-in-place roofing and as facing for panels used mostly on exterior walls.

When it will be factory coated with a finish, the Standard A 875 aluminized steel, which consists of 95 percent zinc and 5 percent aluminum, is often pretreated, using complex oxides, zinc phosphates, or chromates in accordance with ASTM D 2092. The Standard A 792 aluminized steel, which consists of 55 percent aluminum and 45 percent zinc, is often pretreated with chromates.

Other Metal-coated Metals

There are many other kinds of metal-coated metals used in building construction, including the common types that follow.

Cadmium-plated Steel. Steel fasteners and hardware items are sometimes electroplated with cadmium to provide electrolytic separation between the steel and other materials, to prevent galvanic corrosion, which is discussed under "Galvanic Corrosion" in the heading in Chapter 3 entitled "Types of Corrosion." Since cadmium is toxic, cadmium-plated steel items should not be used where they will come in contact with people, especially children, or food. Cadmium-plated steel should comply with the requirements of ASTM Standard A 165.

Chromium-plated Steel. Iron, carbon steel, and stainless steel are all chromium plated to make products used in building. The applicable standards include ASTM standards B 177, B 254, B 320, and B 650.

Chromium-plated Aluminum. Chrome is plated onto aluminum after zincating (see "Zinc-plated Aluminum" later in this section) or plating the aluminum with copper, brass, or nickel.

Chromium-plated Copper. Copper can best be plated with chromium after a coating of nickel has first been applied. Chromium-plated copper has many uses in the construction industry, especially on hardware, toilet accessories, and fasteners.

Nickel-plated Steel. Nickel plating is sometimes used as an intermediate coat beneath chromium or copper plating. It is seldom used alone in the building industry.

Nickel-plated Copper. Nickel plating on copper is used primarily as a base for chromium plating.

Brass-plated Steel. Brass is a copper–zinc alloy (see Chapter 3). Brass-plated steel is sometimes used for builders' hardware and in similar items.

Copper-coated Steel. Copper-coated steel sheets are sometimes used in flashing and other sheet-metal applications.

Copper-plated Aluminum. Copper plating may be used either as a base for chromium plating or as a finish.

Terneplate. Terne is an alloy of lead and tin, usually in a ratio of four parts lead to one part tin. Terneplate, which is the so-called terne used in older buildings, is actually sheet iron or steel coated with a layer of terne. Today's terne-coated steel (TCS) is copper-bearing steel or stainless steel coated with terne. Terneplate and TCS are used mostly in roofing, but they are occasionally found in other nonstructural metal items.

Bare, terne-coated metal to be coated with another material should first be cleaned with mineral spirits, then wiped dry with clean cloths. After this the surfaces should immediately be given a coat of pretreatment solution specifically formulated for use on terne-coated metal. Such pretreatment should contain 95 percent linseed oil. The pretreatment should be allowed to dry for at least 72 hours or longer, if necessary, to ensure that it is thoroughly dry, before applying a primer.

Tinplate. Tin is a bluish-white crystalline metal that is both malleable and ductile. Tinplate is a sheet of steel or iron that has been coated with a layer of tin. Tinplate is now seldom used in commercial buildings, but it may still occur in an existing building. It is used primarily in making food containers, especially tin cans, but may be found in older buildings as roofing.

Lead-coated Copper. Sheets of lead-coated copper are used extensively in roofing and flashing applications and in other nonstructural metal items, especially when having water run across bare copper would be harmful to adjacent materials. Lead-coated copper is produced by electroplating or hot dipping. The lead may be on both sides of the sheet, or on one side only. The standard for lead-coated copper is ASTM Standard B 101, which lists two types and two classes. Type I denotes application of the coating by dipping. A Type II coating is electrodeposited. The Class A Standard requires a lead-coating weight of 6 to 7-1/2 pounds per 100 square feet of surface. The Class B Heavy standard requires 10 to 15 pounds of coating per 100 square feet. These weights give only the weight of the lead on each side of a sheet. Thus, the total weight of the lead on a Class A Standard sheet with lead on both sides, which is the way Standard B 101 lists it, is from 12 to 15 pounds per 100 square feet of sheet.

Zinc-plated Aluminum. In the process called "zincating," aluminum plate is immersed in a zincate bath to coat it with a thin film of zinc. Zincating is generally used to prepare aluminum for electroplating. The standard for zincating on aluminum is ASTM Standard B 253.

Gold Plating (Gilding). Gilding is an ancient art at least four thousand years old that is still practiced today. Gold plating is the application of gold leaf to wood, plaster, metal, glass, or another material to simulate the look of solid gold. It does not include the application of gold paint. There are some imitation golds known as dutch leaf that are used to simulate real gold leaf, but this material tarnishes (oxidizes), even when protected by varnish. Pure gold needs to protection, since gold does not oxidize, for all practical purposes. Even extremely thin high-carat gold leaf will, when properly maintained and protected, withstand most climates and pollution levels for between 30 and 90 years.

The purity of gold is measured by its weight in carats. Pure gold weighs 24 carats. Golds of lesser carat weight are alloys that include silver, copper, or both. Gold leaf can weigh as much as 23-1/2 carats but most is lighter, because pure gold does not wear as well as its lower-carat alloys.

Gold leaf is available today in four forms. Loose leaf comes in loose sheets, for use on interior surfaces only. Patent leaf comes in books of sheets adhered to backing paper. It is the form used for most architectural gilding, especially on exteriors. Surface leaf is used where close inspection is not critical, since it is made from scrap gold and has flaws and discolorations. Finally, glass leaf is used on signs and glass. Among these four forms there are many qualities and grades.

There are essentially three methods of gilding is general use today, although others are also used. These three are oil (mordant) gilding, water gilding, and glass gilding.

Different gilders have their own ways of using each method. In essence, however, both the oil and water methods are started by applying several coats of a material known as gesso, which is a mixture of calcium carbonate (Gilders Whiting) and glue. In oil gilding the gesso is then coated with shellac or marine varnish. The gold leaf is next laid into a coat of oil sizing, which acts as an adhesive. In water gilding, the gold leaf is laid over the gesso in a wet mixture of alcohol and water called gilders' liquor. If the gold leaf is to be burnished, a layer of a special burnishing clay is laid just ahead of the gold. In glass gilding, an oil size is sometimes used, but usually the gold is held in place by a size that is a mixture of gelatin and water.

Other Gilding Metals.   Silver, platinum, and palladium, and other alloys containing them, are sometimes used for gilding, in a manner similar to that used for gold. Gilding with these metals is, however, seldom used in building or related construction.

Anodic Coatings

Anodizing is an electrolytic, anodic oxidation treatment that produces a thicker oxidized coating on a metal than is natural. The coating protects the metal against further oxidization and abrasion. Anodized coatings can also be dyed, producing a colored decorative finish. Many metals can be anodized, but of those metals used in the building industry aluminum has proved to be the only practical one.

The anodizing process consists of immersing aluminum in a tank containing an acid and passing an electric current between the metal and the acid. The oxide resulting on the aluminum may be clear, opaque, or translucent, depending on the chemicals used and the aluminum alloy. The sizes of tanks available limit the sizes of aluminum items that can be anodized.

The thickness of an anodic coating on aluminum can be controlled. Thin anodic coatings, called "flash coatings," are used primarily as a pretreatment for paint or other organic coatings. Anodic coatings that are usually called "anodizing" are much thicker.

Even the thicker anodic coatings seldom provide a complete finish on a piece of aluminum. Usually, a mechanical finish (see the section earlier in this chapter on "Mechanical Finishes for Aluminum") is followed by one of the chemical-etched finishes (see "Chemical Finishes for and Cleaning of Aluminum") before the anodizing process begins. In addition, foreign matter, such as oil, grease, soil, and other contaminants, must be removed before aluminum can be anodized effectively. Any one or several of the chemical-pretreatment types mentioned earlier under "Chemical Finishes for and Cleaning of Aluminum" may be used as necessary, depending on the finish desired and the contaminants present. In addition, anodized aluminum that is not to receive an organic coating must be sealed, to overcome the natural porosity of the anodic coating and protect the underlying metal. The methods generally used are boiling in either pure water or in a nickel-acetate solution.

Types of Anodic Coatings on Aluminum

There are five commonly used processes for anodizing aluminum. The first, the sulfuric acid process, is the only one widely used on aluminum that is used in building and related construction. The other four processes are seldom, if ever, used on aluminum for use in buildings. These are the chromatic acid process, the oxalic acid process, the phosphoric acid process, and the boric acid process.

Colors

Anodizing produced by the sulfuric acid process can be easily treated to produce color-anodized aluminum. Oxalic acid anodizing is also easily colored, but because it is an expensive process it is much less used on aluminum for buildings than that produced by the sulfuric acid process.

Coloring may be done in several ways. A few gold colors are produced by impregnating the anodizing with dyes or pigments. Other colors may be produced in the same way, but they are often not colorfast. Some colors are produced by electrolytically depositing pigments in the anodizing.

Probably the most colorfast and widely used colors are integrally produced by carefully selecting the alloy, chemicals, and methods to be used. Most integral colors are proprietary and must be requested by name.

Standard Designations for Anodic Coatings

The Aluminum Association classifies anodic coatings into four categories: general, protective and decorative, architectural class II, and architectural class I. The first two categories apply to general industrial work and are not used when referring to anodized aluminum products for use in building and related construction.

Architectural Class II.   Coatings in architectural class II must not be less than 0.4 mils in thickness and weigh not less than 17 milligrams per square inch. They range from these lows up to the lowest thickness and weight permitted for architectural class I coatings.

Architectural class II coatings are most often used in locations where they will not be subject to abuse. They are also used for windows and portions of storefronts and curtain walls that are regularly cleaned and not subject to contact by the public. They are divided into the following categories:

Clear (natural) (A41).

Integral color (A42).

Impregnated color (A43).

Electrolytically deposited color (A44).

Other (A4x).

Clear architectural class II anodic coatings are usually produced by the sulfuric acid process.

Integral color anodic coatings are produced in accordance with the users' requirements. Most are proprietary.

Impregnated color anodic coatings are produced by the sulfuric acid process and then dyed.

Electrolytically deposited color coatings are usually produced by the sulfuric acid process and electrolytic deposition of pigments.

The Other category is again reserved for anodic coatings that do not fit into the other classes. When it is used, a full description of the finish must follow.

Architectural Class I.   Coatings in the architectural class I category must weigh at least 27 milligrams per square inch and be at least 0.7 mils thick, but many are much heavier and thicker. For example, some members of a type of anodic coatings called "hardcoat," produced using proprietary alloys and processes, are as much as 3 mils thick. Hardcoat finishes are also harder, denser, and heavier than those produced by other anodizing processes.

Architectural class I anodic coatings are most often used in hard-to-reach exterior locations and in both interior and exterior locations where they will be subject to abuse or contact by the public. They are divided into the following categories:

Clear (natural) (A31).

Integral color (A32).

Impregnated color (A33).

Electrolytically deposited color (A34).

Other (A3x).

Clear architectural class I anodic coatings are usually produced by the sulfuric acid process.

Integral color anodic coatings are produced in accordance with the user's requirements. Most are proprietary. They include the so-called hard-coat finishes.

Impregnated color anodic coatings are produced by the sulfuric acid process, then dyed or colored using mineral pigments.

Electrolytically deposited color coatings are usually produced by the sulfuric acid process and the electrolytic depositing of pigments.

The Other category is reserved for anodic coatings that do not fit into the other classes. When this classification is used, a full description of the finish must follow.

Vitreous Coatings

The only type of vitreous coating on metal that is commonly used in the building trades is porcelain enamel, a form of vitreous organic coating that displays most of the characteristics of glass. For use in buildings it is fused to a backing metal at a high temperature. Materials baked at temperatures of 800 degrees Fahrenheit or higher comply with the ASTM's definition of porcelain enamel, but most of it is fired at temperatures between 1,450 and 1,550 degrees Fahrenheit. For most building-related purposes, porcelain enamel is applied over low-carbon steel or aluminized steel. For other uses it may also be applied over stainless steel and aluminum. It is seldom used over copper alloys for use in buildings.

The types of porcelain enamels used in buildings are hard, abrasion resistant, and nonporous. Water and atmospheric pollutants will barely penetrate them. Their color is as permanent as that of any other material used in building construction. They are also available in a great number of colors, textures, and patterns.

Porcelain Enamel on Steel

Steel to which porcelain enamel is to be applied is either a special material produced for this purpose, decarbonized enameling steel, or conventional cold-rolled sheets.

Almost all porcelain enamel on steel is applied to sheets that are from 14 to 22 gage in thickness before the enamel is applied. Most panels are four feet by eight feet or smaller, although larger panels are possible.

Most applications require two coats of enamel. The first (ground) coat forms a permanent bond with the base metal. The second (top or cover) coat contains the coloring elements and frits that give the porcelain enamel its color, gloss, and corrosion resistance.

Porcelain enamel on steel should comply with the recommendations of the Porcelain Enamel Institute. Their applicable standards include the following:

Guide to Designing with Architectural Porcelain Enamel on Steel.

S-100, "Recommended Specifications for Architectural Porcelain Enamel on Steel for Exterior Use."

"Color Guide for Architectural Porcelain Enamel."

"The Weatherability of Porcelain Enamel."

Bulletin T-2, "Test for Resistance of Porcelain Enamel to Abrasion."

Bulletin T-20, "Image Gloss Test."

Bulletin T-21, "Test for Acid Resistance of Porcelain Enamels."

Bulletin T-22, "Cupric Sulfate Test for Color Retention."

In addition, several ASTM standards are applicable to porcelain enamel on steel, including standards C 282, C 283, C 286, C 313, C 346, C 448, C 538, C 540, and E 97.

Porcelain Enamel on Aluminum

The porcelain enamel used on aluminum is slightly different from that used on steel. It is formulated to fire at lower temperatures and does not always require a ground coat. Some colors and patterns do, however, require that a ground coat be applied.

Only certain alloys and tempers of aluminum are suitable to receive porcelain enamel, because of the effects of the baking process, which tends to heat treat and anneal aluminum.

As is true for steel, almost all porcelain enamel on aluminum is applied to sheets that must be stiff enough to prevent warping during the enameling process. Alternatively, the sheets may be given a coat of porcelain enamel on their concealed sides to balance the stresses imposed by the porcelain enamel on the exposed sides, or they may be reinforced to prevent warping. The sheets must also be smooth and free of die lines that will show through the enamel.

In any case, porcelain enamel on steel should comply with the recommendations of the Porcelain Enamel Institute. Their applicable standards include:

S-105, "Recommended Specifications for Architectural Porcelain Enamel on Aluminum for Exterior Use."

"Color Guide for Architectural Porcelain Enamel."

"The Weatherability of Porcelain Enamel."

Bulletin T-2, "Test for Resistance of Porcelain Enamel to Abrasion."

Bulletin T-20, "Image Gloss Test."
Bulletin T-21, "Test for Acid Resistance of Porcelain Enamels."
Bulletin T-22, "Cupric Sulfate Test for Color Retention."
Bulletin T-51, "Antimony Trichloride Spall Test for Porcelain Enameled Aluminum."

In addition, several ASTM standards are applicable to porcelain enamel on aluminum, including standards C 282, C 283, C 286, C 313, C 346, C 448, C 538, C 540, C 703, and E 97.

Laminated Coatings

Laminated coatings include all the adhesive-bonded plastic coatings used today over sheet metals. The plastic films used in these coatings include, but are not necessarily limited to, polyvinylchloride (PVC) and polyvinyl fluoride (PVF). These materials are laminated to steel, galvanized steel, aluminum given a conversion coating to prepare it, and other nonferrous metals. Sheets with laminated coatings are available for a wide variety of uses, including both interior and exterior applications. One of the most common uses today is for aluminum siding.

Reasons for Mechanical and Inorganic Metal Finish Failure

By definition, nonstructural metal in buildings is always supported by or attached to something. Therefore, some failures of the types of finishes discussed in this chapter can be traced to problems with supporting or underlying materials or systems. The possibilities include one or more of the following: concrete or steel structure failure, concrete or steel structure movement, metal framing and furring problems, wood framing and furring problems, solid substrate problems, and other building-element problems, such as those discussed in Chapter 2 under their headings as listed in the Contents. Many of these causes of metal finish failure, while perhaps not the most probable ones, may be more serious and costly to fix than the types of problems discussed in this chapter. Consequently, the possibility that they may be responsible for a particular metal finish failure should be investigated.

The causes discussed in Chapter 2 should be ruled out or repaired if found to be at fault before metal finish repairs are attempted or a new nonstructural item is installed to replace the failed item. It will do no good to repair a metal finish, apply a new finish, or install a replacement item if a failed, uncorrected, or unaccounted-for problem of a type discussed in Chapter 2 exists: the repaired or new finish or item will also fail.

Other possible causes of finish failure are discussed in Chapter 3 under "Errors That Lead to Corrosion" and "Errors That Lead to Fracture" and their preceding text. Those possible causes should be investigated and ruled out before attempting to repair or refinish a failed finish.

Still more possible causes of metal finish failure are addressed in Chapters 6 and 7 in a number of sections. Refer to "Evidence of Mechanical and Inorganic Finish Failures" later in this chapter for references to the applicable headings in Chapters 6 and 7.

After the problems discussed in Chapters 2, 3, 6, and 7 have been investigated and found to be not present or have been repaired if found, the next step is to discover any additional causes for the mechanical or inorganic metal finish failures and correct them. Such causes include bad materials, metal failure, improper design, bad workmanship, failure to protect the installation, poor maintenance procedures, and natural aging. The following discussion covers these additional causes. Refer to "Evidence of Mechanical and Inorganic Finish Failures" later in this chapter for a listing of the types of failures to which the causes addressed here apply.

Bad Materials

There are several ways in which the materials used may have been defective. For example, the underlying metal may have been improperly produced, or a coating material may have been incorrectly manufactured. Improperly produced metals or finishing products are certainly not unheard of, so this possibility should be considered when a metal finish fails. The number of incidents of bad materials, however, is small compared with the number of cases of bad design or workmanship. The following sections cover some of the types of manufacturing defects that might occur.

Incorrectly Produced Metal Materials.   A steel may have the wrong carbon content, for example, or incorrect amounts of constituents in an alloy, or the metal may have been improperly quenched or annealed. There are many possibilities. Improperly manufactured metal, however, is not a significant cause of finish failure.

Bad Coating Material.   A coating material with inconsistent color, composition, or density may have been used. An improper coating formulation or defective or old materials may prevent the coating material from curing properly or developing the characteristics expected.

Metal Failure

Corrosion or fracture in a finished metal, which are discussed in detail in Chapter 3, will ultimately result in finish failure. Because mechanical finishes are a part of the metal itself, the effect on them of corrosion or fracture is the same as on the underlying metal. Many applied finishes, such as galvanizing and aluminizing, will also be damaged immediately by fracture. Corrosion of an underlying metal with an applied finish does not often occur, though, unless there has been previous damage to the applied finish. When a metal cracks, most coatings applied to it will also crack. There are exceptions, however. Some coatings, especially the laminated ones, may withstand initial cracking and not show damage until the damage to the underlying metal is quite advanced. Some types of damage to the underlying metal that may lead to finish failure include those that follow.

Corrosion Damage.   It is often difficult to determine if a fracture has permitted corrosion to begin or corrosion has caused the fracture to form. The two are usually linked in some way. Refer to "Errors That Lead to Corrosion" and "Errors That Lead to Fracture" and the pages that precede them in Chapter 3 for a detailed discussion. If steel rusts beneath a coating the coating will fail, whether or not there is additional damage (fracture) to the steel caused by the corrosion.

Fracture.   Refer to "Errors That Lead to Fracture" in Chapter 3 for a discussion of fracture.

Improper Design

Improper design includes selecting the wrong paint or transparent finish material for the location and conditions as well as requiring an improper installation. The design problems discussed below can lead to mechanical or inorganic finish failure.

Inappropriate Finishes

Selecting an inappropriate finish for use in nonstructural metal items can lead to failure. For example, neither as fabricated (mill finish) aluminum nor as fabricated copper-alloy items should be used in locations where appearance is important, because they are likely to contain flaws and discolorations.

The No. 2D, 2B, and bright rolled stainless steel mill finishes, as well as proprietary finishes on any metal, which cannot be matched in a fabricator's shop, should not be required as the final finish for metals that must be fabricated into nonstructural items of the type addressed in this book. An exception to this rule might be made where appearance is unimportant, however.

Mill-induced discolorations and imperfections are often made to appear even worse by welding and by bending or other forming of the metal.

Requiring that a mechanical finish be used on clad sheet, such as aluminum-clad aluminum, will often result in having the finishing process penetrate the thin cladding, thus defeating the purpose of the cladding.

Requiring that a buffed or other highly polished finish be applied on metals to be used on large flat surfaces will likely produce a rippling effect known as "oil canning." Flat, unpatterned metals are most likely to show this effect. The use of nonreflective matte, textured, patterned, or etched (embossed) sheet metals will usually reduce—and often eliminate—this effect.

Requiring that thin metal sheets have an abrasive blasted finish will often lead to thinning of the sheet metal to the extent that it will buckle or fracture when subjected to even small stresses. Blasting may also cause thin sheets to distort.

To require that aluminum on a large flat surface have a chemically brightened finish will probably produce an unsatisfactory finish, because chemically brightened surfaces are difficult to make uniform on large surfaces.

Using matte-dipped or bright-dipped chemical finishes as final finishes on copper-alloy items does not usually prove satisfactory. These finishes are hard to control and do not usually present uniform surfaces when used alone.

Requiring that wire-brushed finishes be used on large surfaces is not a good idea, because they are difficult to maintain. Normal maintenance will sometimes dull them or make them appear nonuniform.

An anodic coating that is too thin for the use intended will soon wear and appear unsightly. For example, an architectural class II coating on main-entrance lobby doors will soon wear away due to constant use.

A zinc coating too thin for the exposure to be expected will soon corrode so that it no longer protects the underlying metal.

Selecting the wrong metal alloy to use as a base for a porcelain-enamel finish will lead to failure of the finish. Fortunately, such a failure will usually appear immediately.

Using imitation gold for gilding will not produce the desired result if gilding is really what is desired. Imitation gold will not last as long or look as good, even initially.

Mismatched Finishes

Failing to match properly the finishes used on adjacent components in a metal fabrication will produce an unsightly condition. If, for example, a chemical finish is used on aluminum assembly components made from different alloys, the components are not likely to match in appearance. Each alloy is likely to look different, even when identical chemicals and processes are used.

The different aluminum alloys, anodic colors, and procedures to be used to produce a finish on each component of an anodized aluminum assembly must be selected carefully to ensure that the various components exhibit the same appearance.

Copper alloys in the same fabrication must also be carefully selected for compatibility. Different copper alloys produce varying finishes, even when the same process and chemicals are used. Refer to the CDA publication *Copper, Brass, Bronze Design Handbook: Architectural Applications* for guidance.

Requiring that butt joints be used in large anodized aluminum or copper-alloy panels will produce an unsatisfactory appearance. No two anodized panels or finished copper-alloy panels will ever match exactly in color and shade.

Trying to match one type of finish with another is seldom successful. For instance, a color-anodized aluminum member will almost surely not match a member finished with a coating of any kind, whether it is inorganic or organic. Producers sometimes claim that such a match is available, but the author has yet to see one that is successful. Mismatching is not always the case, however, when both materials are organic coatings. Refer to Chapter 5 for additional information.

Improper Preparation for Finishing

The proper preparation of a surface that is to receive a finish is critical to the success of the finish. The errors now discussed will often lead to finish failure.

The most common error is that of failing to require the proper preparation of surfaces to receive their finishes. One example might be failing to require the chemical precleaning of metal.

Failing to require the use of either mechanical or chemical finishes or both under an anodic coating on aluminum sheet material or to use an embossed sheet will lead to an unsatisfactory finish. Anodic coatings on large, flat sheet-aluminum surfaces that have not been mechanically or chemically finished to roughen them often exhibit streaks or discolored areas that would have been concealed by an underlying texture.

The failure to require proper pretreatment beneath a metallic coating will often result in delamination of the metal coating. Failing to require a layer of nickel plate between chromium plating and underlying copper is an example. Another is failing to require zincating on aluminum that is to receive copper plating.

A common error is failing to require proper treatment and protection for metallic-coated steel. There is a tendency, for example, to believe that galvanized steel is immune from corrosion, but this is far from the truth. The zinc coating is itself subject to white rust, which will interfere with paint adhesion. The zinc coating on galvanized metals is also subject to galvanic corrosion from more noble metals, such as copper, and to chemical corrosion from soluble sulfites like those found in cinders. Zinc will also be corroded by salt, some concrete aggregates, the acids in cedar and oak, and acid rain. The zinc coating itself may oxidize or fail to protect the material's ferrous metal core when it is submerged in water or subjected to heavy layers of condensation or other water deposits. Zinc coatings will also be burned away during welding. In addition, zinc coatings are actually expected to corrode. One of their strong points is that they are sacrificial in nature, meaning that they corrode in lieu of the coated steel. Only when a significant amount of the zinc coating has corroded away will the steel begin to corrode. This sacrificial action protects the steel far beyond the normal protection that might be expected from the coating alone. Eventually, though, all unprotected galvanizing will fail (Fig. 4-4).

Bad Workmanship

Correct preparation and installation are essential if metal finish failures are to be prevented. The following workmanship problems can lead to such failures.

The primary error is usually that of not following the design and recommendations of the finish manufacturer and the standards generally recommended by the trade or association responsible for setting standards for the materials and process being used. Applying finishes under conditions of temperature, humidity, or other conditions different from those recommended or otherwise improperly applying a mechanical, chemical, or other finish are examples.

Another error is failing to remove anodizing acids from anodized aluminum. If the acid used to produce anodized aluminum is not properly removed from the aluminum, it will eventually wash off and damage the aluminum and adjacent surfaces. Acids trapped in hollow sections are a particular problem, as are acids trapped in joints between assembled sections. Weep holes are usually needed in hollow sections and assemblies to ensure the proper removal of acids.

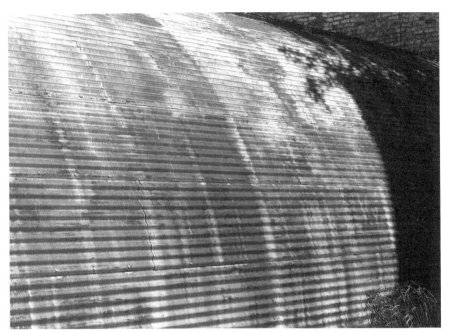

Figure 4-4 The galvanizing on this old quonset hut has finally given way to time and atmospheric pollution. The dark areas are rust just beginning to show through the zinc coating.

Applying a finish over an improperly applied pretreatment or underlying finish will usually lead to failure of the finish. Nonadherence or other failure of an underlying pretreatment or finish may cause the next coating to fail. This can be a problem regardless of the type of pretreatment or underlying finish that has failed, but it is most likely to happen when an underlying metallic coating fails. If the zincating on aluminum does not adhere, for example, an applied chromium plating may delaminate.

Failing to clean an underlying metal properly before applying a finish will usually cause the finish to fail. Grease and mill scale must be removed from aluminum, for instance, before an anodic coating is applied. Mill scale, dirt, oil, grease, and even fingerprints must be removed from copper alloys before a conversion coating is applied to produce a patina or statuary bronze finish. Such contaminants will mar the finished surface and may even appear worse after the finishing than before.

Another error is failing to remove strippable coatings soon after installation. Such coatings may become difficult if not impossible to remove after prolonged exposure to sunlight.

Applying a porcelain-enamel finish to items with inside or outside corners that are too sharp will lead to failure of the finish. The minimum corner radius should be not less than 3/16 inch in steel or 1/16 inch in aluminum. Improperly applied gold leaf will not adhere properly.

Failure to Protect the Finish

Metal finishes must be protected before, during, and after installation. Some of the errors found include those that follow.

Failing to protect mechanical finishes from damage during subsequent operations can lead to scratched, eroded, or otherwise damaged finishes.

Contaminants, including airborne dust and other solid particles, that are permitted to fall into chemicals during finishing, to interfere with mechanical finishing operations, or to fall onto wet, newly finished surfaces will affect the final appearance of a finish and even cause it to fail.

Permitting abuse before or during application or after a finish has been applied may be the greatest single cause of damaged finishes. For example, stacking galvanized steel items in such a way that moisture is allowed to accumulate on them will lead to early corrosion. Permitting contact by vehicles may also be a major problem, as demonstrated by Figures 4-5 through 4-7.

Another cause of damaged finishes is that of permitting adjacent materials to be cleaned with cleaning materials that will harm the metal. An often-occurring problem is the cleaning of adjacent masonry by acid cleaners. If this is permitted, damaging of metal due to runoffs or splattering is almost inevitable, even when the masonry cleaner attempts to take precautions to prevent contact between cleaner and metal. Acid-type cleaning materials should never be permitted in the vicinity of metals under any circumstances.

Poor Maintenance Procedures

Not maintaining metal finishes properly can result in their failure. Poor maintenance procedures include those that follow.

Failing to remove grease, dirt, and other contaminants regularly can result in permanent stains.

The failure to repair promptly damage to finishes on metals (see Fig. 4-5) can lead to more serious damage later.

Improperly cleaning anodized aluminum can damage its finish. Cleaners must be neutral, because both alkaline and acid materials will damage the coating. Ordinary wax cleaners and mild soap and detergents are the strongest substances that should be used without consulting the metal and finish producers (see "Cleaning and Maintaining Existing Metal Surfaces" later in this chapter).

Figure 4-5 A truck did the initial damage here, but failure to repair it led eventually to loss of the organic coating and eventual corrosion of the underlying galvanizing and rusting of the steel base sheet.

Figure 4-6 A truck inside the warehouse backed into the metal wall panels, resulting in the bowed-out panels shown.

Figure 4-7 The damaged finish on the door jamb shown in this photo resulted from a combination of using the wrong material for the installation (a sheet metal on an industrial door jamb) and failing to protect the jamb from damage after it was installed.

Failing to reoil statuary bronze, which is necessary to maintain the appearance of its finish, is another maintenance problem that can lead to failure. Refer to "Cleaning and Maintaining Existing Metal Surfaces" later in this chapter.

Metal finishes may also be damaged severely if cleaned using sandpaper, steel wool, and abrasive or caustic cleaners.

Improper or premature cleaning of copper alloys can also lead to unwarranted damage. Patinas and statuary bronze finishes may be removed or at least severely damaged by improper cleaning.

Applying unnecessary coatings in an attempt to protect copper alloys often damages them in the process.

Improperly cleaning gold leaf can remove the gold leaf. Different cleaning methods are required, depending on the type of gilding used.

Natural Aging

All metal finishes age and lose at least some of their properties over time (see Fig. 4-4). The finish producer or an association representing the producer is the best source of data about the expected life of a finish. It should be kept in mind, however, that the producer's projections may assume optimal conditions.

Metal finishes may fail for any number of reasons. Failing to recoat or refinish a material at proper intervals can lead to premature aging and failure. Most applied metal finishes that are exposed to the sun's ultraviolet rays and weathering will fail faster than those used in other locations. In any case, all applied finishes will fail eventually. Failure will occur sooner, however, if finishes are not properly maintained and recoated or reapplied at appropriate intervals.

Old laminated coatings may lose their elasticity and no longer be able to respond to the expansion and contraction of the material to which they are applied. This may result in cracking or delamination.

Weathering will slowly wear away finishes until they eventually lose their appearance or do not protect metal properly. For this reason anodic coatings on aluminum must be carefully selected so that surfaces to be exposed are heavier and more corrosion resistant than those used in less-exposed locations.

Evidence of Mechanical and Inorganic Finish Failures

In the following discussion mechanical and inorganic metal finish failures are divided into types of failure such as ''Damaged or Discolored Finishes.'' Each type of failure is listed under its own heading. The list following each type indicates the location in this book where the causes of that type of failure are to be found.

A discussion of the errors that can lead to each failure cause listed can be found in the chapter whose number follows the listing, under a heading with the same name as that of the failure. To make each cause of failure easier to locate, each is included in the Contents.

Thus, to discover a possible cause for a damaged finish on a metal fabrication, look under the heading ''Damaged or Discolored Finishes.' One possible cause listed there is ''Other Building Element Problems: Chapter 2.'' Open Chapter 2 to ''Other Building Element Problems.'' Examine the problems discussed there that can lead to failure, to discover if any of them may be applicable to the type of damage observed. If none of those reasons seem to correlate with the damage, look under the next possible cause and so on, until the actual cause is identified. Of course, there may be more than one cause for a particular failure.

Rust or Other Corrosion

The appearance of rust or other corrosion on a metal's surface may have been caused by many factors. Most of them are addressed under one or more of the following headings:

- Other Building Element Problems: Chapter 2.
- Errors That Lead to Corrosion: Chapter 3.
- Bad Materials: Chapter 4.
- Metal Failure: Chapter 4.
- Improper Design: Chapter 4.
- Inappropriate Finishes: Chapter 4.
- Mismatched Finishes: Chapter 4.
- Improper Preparation for Finishing: Chapter 4.
- Bad Workmanship: Chapter 4.
- Failing to Protect the Finish: Chapter 4.
- Poor Maintenance Procedures: Chapter 4.
- Natural Aging: Chapter 4.
- Improper Design: Chapter 6.
- Improper Fabrication: Chapter 6.
- Bad Installation Workmanship: Chapter 7.
- Failure to Protect the Installation: Chapter 7.

Damaged or Discolored Finishes

Metal finishes can be damaged in many ways and at many levels, from minor scratches to the complete delamination of an applied finish. Unfortunately, there is no easy way to determine the specific cause of some finish failures, though others will be obvious at a glance. A stain from a water leak will be obvious, for example, but a cracked applied finish may be the result of causes as diverse as bent metal or errors in applying the finish.

Discolorations may actually be stains—or simply the result of selecting the wrong finish initially. Some apparent discolorations may not be discolorations at all, but rather the result of using the same finish on different alloys of the same metal.

Depending on the finish and the type of damage or discoloration to it, a finish failure may be due to a failure addressed under one or more of the following headings:

- Concrete and Steel Structure Failure: Chapter 2.
- Concrete and Steel Structure Movement: Chapter 2.

- Metal Framing and Furring Problems: Chapter 2.
- Wood Framing and Furring Problems: Chapter 2.
- Solid Substrate Problems: Chapter 2.
- Other Building Element Problems: Chapter 2.
- Errors That Lead to Corrosion: Chapter 3.
- Errors That Lead to Fracture: Chapter 3.
- Bad Materials: Chapter 4.
- Metal Failure: Chapter 4.
- Improper Design: Chapter 4.
- Inappropriate Finishes: Chapter 4.
- Mismatched Finishes: Chapter 4.
- Improper Preparation for Finishing: Chapter 4.
- Bad Workmanship: Chapter 4.
- Failing to Protect the Finish: Chapter 4.
- Poor Maintenance Procedures: Chapter 4.
- Natural Aging: Chapter 4.
- Improper Design: Chapter 6.
- Improper Fabrication: Chapter 6.
- Bad Installation Workmanship: Chapter 7.
- Failing to Protect the Installation: Chapter 7.

Breaks and Cracks

There are many possible errors that can lead to breaks and cracks in a metal's finish. Some are related to the metal itself. Many are discussed and identified in Chapter 3 under "Errors That Lead to Fracture," listed in the Contents. The material immediately preceding that heading describes the nature and various types of fracture that can occur in metals. In addition, errors discussed under the following headings may be responsible for breaks and cracks in a finish:

- Concrete and Steel Structure Failure: Chapter 2.
- Concrete and Steel Structure Movement: Chapter 2.
- Metal Framing and Furring Problems: Chapter 2.
- Wood Framing and Furring Problems: Chapter 2.
- Solid Substrate Problems: Chapter 2.
- Errors That Lead to Corrosion: Chapter 3.
- Errors That Lead to Fracture: Chapter 3.
- Bad Materials: Chapter 4.

- Metal Failure: Chapter 4.
- Improper Design: Chapter 4.
- Inappropriate Finishes: Chapter 4.
- Improper Preparation for Finishing: Chapter 4.
- Bad Workmanship: Chapter 4.
- Poor Maintenance Procedures: Chapter 4.
- Natural Aging: Chapter 4.
- Improper Design: Chapter 6.
- Improper Fabrication: Chapter 6.
- Bad Installation Workmanship: Chapter 7.
- Failing to Protect the Installation: Chapter 7.

Worn or Missing Finishes

All finishes will eventually wear away, but erosion may be hastened by certain causes. Many of these will be found under the following headings:

- Errors That Lead to Corrosion: Chapter 3.
- Bad Materials: Chapter 4.
- Improper Design: Chapter 4.
- Improper Preparation for Finishing: Chapter 4.
- Bad Workmanship: Chapter 4.
- Failing to Protect the Finish: Chapter 4.
- Poor Maintenance Procedures: Chapter 4.
- Natural Aging: Chapter 4.
- Improper Design: Chapter 6.
- Improper Fabrication: Chapter 6.
- Bad Installation Workmanship: Chapter 7.
- Failing to Protect the Installation: Chapter 7.

Cleaning and Maintaining Existing Metal Surfaces

Metals with mechanical or inorganic finishes must be cleaned and maintained periodically if finish failure is to be prevented. Black marks and other soiling, mortar, sealants, stains, dirt, dust, grease, oil, efflorescence, salts, unwanted paint, and other unsightly contaminants should be removed completely. The means, materials, and methods that follow are general in nature and do not necessarily apply in every situation. In every case, though, the soaps, detergents, cleaning compounds, and methods used should be compatible with the surfaces to be cleaned and adjacent surfaces, so as not to damage them in any way. The materials and methods used should be those recommended by the metal's producer, the fabricator, the finisher, the manufacturer of the finishing material, and applicable industry standards. In no case should a suggestion made here supersede those recommendations. Finishes damaged from use of incorrect products or improper use of cleaning agents should be refinished.

In some cities there are organizations set up specifically to clean buildings. When the metal to be cleaned entails major surfaces of an entire building and such an organization is available, it might be advisable at least to talk to them about it. The organization's references should be carefully checked, of course, and the methods they propose to use should be ascertained and verified with the recommendations of the material's producer.

To prevent streaking, metal surfaces should not be cleaned with bright sunlight on the surface. Cleaning metals in the shade, or on cloudy days, is best. Cleaning should always proceed from top to bottom so that runoff is removed during subsequent cleaning.

Dirt, soot, pollution, cobwebs, insect cocoons, and the like can often be removed from exterior surfaces with a water spray, followed by scrubbing with a solution of mild soap or household detergent and water. The cleaned surfaces should be rinsed thoroughly with clean water and permitted to dry.

Initially, stainless steel, aluminum, and copper alloys should be cleaned using clear water and mild soap. Where necessary they may be washed with mineral spirits, then with mild soap and clean water, and finally rinsed with water and allowed to dry. Abrasives or caustic or acidic cleaning agents should not be used. If these methods do not satisfactorily clean the surfaces, it may be necessary to use more drastic cleaning methods, but only those methods recommended by the metal's manufacturer or the association representing the manufacturer should be used.

Deposits on stainless steel that are particularly difficult to remove may be taken off with phosphoric or oxalic acid compounds. Such materials should be used while strictly following the directions of the manufacturers of the cleaning agent and metal material.

With the concurrence of the metal and finish producers, solvent- or emulsion-based cleaners may be used on aluminum to remove oil and grease. Again, only when the manufacturer and finisher agree should abrasive cleaners be used on aluminum to remove materials that none of the milder methods will remove. Such cleaners often contain oil, wax, and silicones. The cleaning agents in them may be soap, acids or alkalines, and abrasives. On severely damaged or corroded aluminum finishes professionals may also use other cleaning methods, including etching cleaners, steel-wool polishing, power-driven wire-brush cleaners, and steam cleaners. Most of these methods remove all or part of the existing finish, so they should be used only when absolutely necessary. Finishes damaged by these cleaning operations must be renewed. The cleaning materials used must also be removed. Steel-wool particles will, if left on aluminum, rust and cause stains. It is best to use stainless steel wool when steel wool use is appropriate. In some cases a finish may be renewed in the field, but damaged anodic surfaces may be reanodized only by removing the element and shipping it to the anodizing plant.

Aluminum surfaces are sometimes waxed to provide some weather protection and make them easier to clean. Such wax coatings tend to wear away and yellow with time and must be renewed. It is often necessary to remove the old wax buildup before rewaxing the surface.

There are some specialty sealers on the market that may be used to seal anodized surfaces, to restore their original luster and make them easier to clean. These materials are usually applied on older surfaces that have begun to lose their luster or become porous.

Aluminum is also often protected with tape, a clear strippable coating, or a clear organic coating, which is usually lacquer. Refer to Chapter 5 for a discussion of clear coatings on aluminum. The clear coatings referred to there are not the type of strippable coatings sometimes applied to aluminum products to protect them during shipment and installation. Such coatings and tapes should be removed immediately after the immediate danger of damage has passed.

Copper alloys are frequently oiled or waxed to enhance their appearance and help protect them by excluding moisture. The oil or wax will dissipate over time and must be renewed periodically. Linseed oil has traditionally been used to coat copper alloys. Unfortunately, however, boiled linseed oil contains varnish, which weathers unevenly and may produce a yellow coating and a mottled appearance. Raw linseed oil dries slowly and tends to pick up dust and other contaminants while drying. It also reacts with copper, which partially defeats the purpose of oiling it. Lemon, lemon grass, castor, and other oils are also used, with differing degrees of success. The Copper Development Association's current literature suggests that high-grade paraffin oils may be the best material to use in many situations.

The waxes used on copper alloys include mixtures of Carnauba wax or beeswax with wood turpentine and high-quality commercial paste waxes. They are used mostly in architectural components that are accessible to people. Bronze sculptures intended for people to touch or climb on are often waxed, for example, as are handrails and similar components.

Bronze is sometimes cleaned by a process called "glass bead peening," which is similar to other abrasive blasting procedures, except that this technique uses small glass beads as the abrasive.

Anyone responsible for maintaining copper-alloy surfaces should contact the Copper Development Association at the address listed in the Appendix on Data Sources for recommendations on the particular situation and copper alloy involved.

The cleaning of porcelain enamel beyond simple washing with mild soap and water should be done only in accordance with the enamel producer's recommendations (see "Where to Get More Information").

Refer to Chapter 5 for information about cleaning organic coatings.

Repairing Failed Mechanical and Inorganic Finishes

The extent of the repairs needed in failed mechanical and inorganic finishes depends on the type and extent of the damage. Repairing such finishes with the item left in place is usually restricted to handling minor damage only. Because all the finishes discussed in this chapter can be applied only in a shop (not with the metal installed in a building), major damage often requires that the metal item with the damaged finish be removed and refinished in the shop. Alternatively, of course, the damaged finish may be removed in the field or properly treated there, and a new finish applied, usually an organic paint coating. Refer to Chapter 5 for a discussion on applying paint over another metal finish.

The following discussion contains suggestions for repairing mechanical and inorganic finishes. Because these suggestions are meant to apply in many situations, they might not apply in a specific case. There will also be many possible cases not specifically covered here. When a condition arises in the field that is not addressed here, advice should be sought from the manufacturer and finisher of the metal. Sometimes it will be necessary to obtain professional help (see Chapter 1). Under no circumstances should the specific recommendations in this chapter be followed without careful investigation and the application of professional expertise and judgment.

Before an attempt is made to repair existing mechanical or inorganic finishes, the metal material's manufacturer and the finish material's manufacturer and applicator should be contacted to obtain their recommendations. In the absence of these recommendations the same advice should be sought from the association representing the manufacturers or finishers of the metals involved. Repairs should then be made in strict accordance with their recommendations. Only experienced workers should be used to repair mechanical or inorganic metals' finishes.

Areas where repairs are to be made should be inspected carefully to verify that the existing materials are as expected and that the repair can be satisfactorily accomplished. This inspection should be made preferably by a representative of the metal manufacturer or finisher or both.

It is difficult and sometimes impossible actually to repair mechanical or inorganic finishes in the true sense of the word. Small patches and repairs will usually appear obvious and will often not be acceptable for that reason. One exception is damaged galvanizing on steel. Unrepaired damage to a zinc coating on steel will result eventually in corrosion of the steel. Damage to a galvanized metal surface should, then, be repaired as soon as possible. Such damage can be repaired with a galvanizing repair paint that should have a high zinc dust content and be formulated specifically for regalvanizing steel. The paint chosen should comply with the Military Specification MIL-P-21035 (Ships). To affect repairs, clean field welds, bolted connections, and abraded areas and scratches, and apply two coats of galvanizing repair paint.

When a metal's mechanical or inorganic finish is in need of repair, it is usually necessary first to remove applied wax, oil, or organic coating. An anodized aluminum item may, for example, have been coated with a lacquer, which must be removed before the aluminum or its anodizing can be repaired. Such coatings are often reapplied after repairs have been completed. Waxes and oil coatings were discussed earlier in this chapter. Organic coatings are discussed in Chapter 5.

The repair of porcelain enamel is possible by using a synthetic porcelain glaze developed specifically for the purpose. This is, however, a job for a professional. Porcelain enamel may also be reenameled. See "Where to Get More Information," below, for sources of additional information.

Where to Get More Information

Probably the best source of information available today about metal finishes is the National Association of Architectural Metal Manufacturers' *Metal Finishes Manual for Architectural and Metal Products*. Anyone interested in designing, applying, or maintaining metals or their finishes should have access to a copy of the latest edition.

Harold B. Olin, John L. Schmidt, and Walter H. Lewis's 1983 edition of *Construction Principles, Materials, and Methods* is a good source of data about the iron, steel, and aluminum finishes addressed in this chapter. Unfortunately, there is little in this book that will help with repairing such finishes.

The following publications from the Copper Development Association include data pertinent to the subjects discussed in this chapter:

- *Copper, Brass, Bronze Design Handbook: Architectural Applications.*
- *Copper, Brass, Bronze Design Handbook: Sheet Copper Applications.*
- "How to Apply Statuary and Patina Finishes."

The following publications from the American Architectural Manufacturers Association contain valuable information about metal finishes:

- "Voluntary Guide Specifications and Inspection Methods for Integral Color Anodic Finishes for Architectural Aluminum."
- "Voluntary Specifications for Residential Color Anodic Finishes."
- "Voluntary Guide Specifications and Inspection Methods for Clear Anodic Finishes for Architectural Aluminum."
- "Voluntary Guide Specifications and Inspection Methods for Electrolytically Deposited Color Anodic Finishes for Architectural Aluminum."
- "Voluntary Guide Specifications for Cleaning and Maintenance of Architectural Anodized Aluminum."

The 1982 American Society for Metals' *Metals Handbook: Ninth Edition, Volume V: Surface Cleaning, Finishing, and Coating* contains detailed, comprehensive discussions of materials and methods to prepare for and apply the finishes discussed in this chapter. It is an excellent source of data for anyone responsible for dealing with finishes used on metals.

The American Iron and Steel Institute's *Stainless Steel: Suggested Practices for Roofing, Flashing, Copings, Fascias, Gravel Stops, and Drainage* offers many details about stainless steel sheet-metal work that are applicable to the kinds of metal items addressed in this book.

Maurice R. Petersen's 1984 series of articles "Finishes on Metals, A View from the Field" discusses mechanical, chemical, and other inorganic finishes and their repair.

Stanley Robertson's 1989 article "Restoring the Gilded Surface" is a good source of further information about gilding with gold leaf and repairing and maintaining gilding. Other sources of such data are listed in this book's Bibliography and identified with a [4].

Until recently, the Zinc Institute was a source for data about zinc in the construction industry. However, a recent call to its number listed in the Appendix, netted a recorded message that the Zinc Institute was no longer in operation and questions should be addressed to zinc suppliers. In addition, the American Anodizers Council (AAC) has produced standards for anodizing that are available upon request. Alice Koller's 1981 article "Hot Dip Galvanizing: How and When to Use It" is also excellent on its subject.

Martin E. Weaver's 1989 article "Caring for Bronze" is a valuable discussion on corrosion, cleaning, and protection of copper alloys. Anyone interested in caring for bronze should obtain a copy through *The Construction Specifier*.

The Porcelain Enamel Institute (PEI) offers no specific advice about the cleaning, repairing, or re-enameling of porcelain enamel, but it will furnish a list of enameling companies that can clean, repair, or re-enamel damaged porcelain-enamel surfaces. It may be more convenient, though, to call a local contractor who handles porcelain-enamel products or a local member of PEI and ask them for the name of a nearby organization that can clean or repair such surfaces.

The AIA *Masterspec* specifications that are followed in the Bibliography by a [4] include data on finishes discussed in this chapter.

The ASTM standards related to the subjects discussed in this chapter are identified in the Bibliography by a [4].

See also the other entries in the Bibliography that are marked with a [4].

CHAPTER
5

Organic Coatings for Metals

The finishes used on nonstructural metal items used in building and related construction are divided in the industry into three broad categories: mechanical finishes, chemical finishes, and coatings. Mechanical and chemical finishes were discussed in Chapter 4.

Coatings are applied finishes that fall into two categories: inorganic and organic coatings. Inorganic coatings were discussed in Chapter 4, along with oils and waxes used to protect the finishes on such metals as aluminum and copper alloys. This chapter discusses organic coatings, including both organic and resinous paint, lacquer, and enamel; clear coatings used to protect the finishes on aluminum and copper alloys; and other organic coatings such as fluorocarbon polymers. Included are the materials, their application, possible failures that may occur in them, and the methods the industry recommends for repairing those finishes and applying them over existing substrates where appropriate.

Chapter 3 discussed the metals that the organic coatings in this chapter may be applied over.

The terms used in this chapter have the meanings listed for them in the Glossary.

Organic coatings are used both to protect and to decorate metals. Because the number of different organic coatings used on metals is enormous and is constantly growing, it is possible here to cover only the ones in most common use at this time (mid 1989) for which information is readily available. The inadvertent omission of a particular type of organic coating does not imply a judgment on the part of the author or publisher as to the value or incidence of that material. We believe that most of the principles presented here, however, also apply to other organic finishes. Under any circumstances the reader is expected to verify the statements made in this book with knowledgeable sources, particularly the manufacturer of the organic coating at issue, before attempting to repair or recoat a given surface.

In general, even though someone knowledgeable about such finishes might guess the type of an existing organic coating material from field examination, positive identification can be made only by taking samples from the material and examining them in a laboratory. In dealing with a failed existing organic coating one should first learn as much as is reasonable about the failed material by reading the information here and in the pertinent references listed in "Where to Get More Information" at the end of this chapter, then consult the manufacturer of the material. When a specific organic coating type is encountered that is not discussed here or in the references mentioned, the next step should be to contact the manufacturer for advice.

Organic Coating Materials

The National Association of Architectural Metal Manfacturers (NAAMM) divides organic coatings into four categories: paints, enamels, varnishes, and lacquers. Thus, according to the NAAMM, paint is a coating, which certainly seems to make sense. Unfortunately, however, the industry as a whole does not universally use the NAAMM's designations. Some coatings (paint) manufacturers and other industry sources, for example, use the terms "paint" and "coating" interchangeably. Others call all their products coatings, and still others call some products paints and others coatings.

Enamels and paints are placed in separate categories by some producers, but not by others, and some call both paint and enamel "paint." There is no generally accepted consensus that paint is just another coating. *Masterspec*, for instance, separates paint and coatings into two different specifications sections. This blurring is somewhat understandable, since coatings technology has undergone such rapid, continual expansion in recent years. There used to be definite distinctions between paints and enamels, and between lacquers and varnishes, but chemistry has changed all that. Even as early as 1969 the NAAMM was saying in its *Metal Finishes Manual* that there was no real distinction anymore between a paint and an enamel. The confusion can be demonstrated just by asking a coatings (paint) manufacturer whether their polyurethane product is a lacquer or a varnish.

There is no intention in this book to quarrel with any source concerning the definitions of the various materials grouped together here as organic coatings. Nevertheless, because the industry's lack of generally accepted definitions makes discussion of the subject sometimes confusing, we have defined in the Glossary some of the terms that are used here. The reader should keep in mind, however, that although an attempt has been made to be reasonable, the definitions in the Glossary are not necessarily those used universally in the coatings (paint) industry.

Manufacturers and Products

Paint and other organic coatings are made and distributed by national, regional, and local manufacturers. Because specific manufacturers are not mentioned in this book, refer to "Where to Get More Information," at the end of this chapter, for more information about finding manufacturers.

There is such a bewildering array of available paint and other organic coatings that it is difficult for anyone who is not a coatings professional to determine which to use in a particular situation. In addition, there are few recognized industry standards for paint or other organic coatings. The only widely recognized standards for many such coatings are those contained in federal specifications, which are unfortunately often out of date, inapplicable to a particular situation, and in some cases even require products no manufacturer offers. Many manufacturers are capable of producing almost any organic coating formulation, but most will not do so unless the amount ordered is enormous. Specially formulated products are often expensive compared to standard available products and will probably not have significantly superior characteristics.

The standard for high-performance coatings, such as fluorocarbon polymers, is the American Architectural Manufacturers Association's "Voluntary Specifications for High Performance Organic Coatings on Architectural Extrusions and Panels" (AAMA 605.2-85). Fluorocarbon polymer coatings and any material sold in competition with them should be required to comply with the latest version of AAMA 605.2-85.

Then AAMA also publishes a performance standard for organic coatings on aluminum, which is called "Voluntary Performance Requirements and Test Procedures for Pigmented Organic Coatings on Extruded Aluminum" (AAMA 603.8-85).

There are hundreds of ASTM standards relative to paints and other organic coatings, but most of them cover only component materials. Because any given paint or other organic coating product may contain many different chemicals, using ASTM standards to control organic-coating products is difficult and impractical.

In addition, where environmental control laws and other legal restrictions are in force, they can drastically affect organic coating selection. Currently (1989) twenty-two states and the District of Columbia have enacted restrictive legislation regarding solvent emission. Some local jurisdictions have followed suit, in some cases passing even stricter legislation than the states. Such laws are eventually expected to appear in other states and jurisdictions.

To make a complicated situation even worse, industry members often do not speak the same language. As noted, a material that some call a paint others may call a coating. And the same quality designation, such as finest, best, and so on, may have entirely different meanings from one manufacturer to another.

Those of us who are not chemists or paint experts must locate and rely on reputable manufacturers and their representatives for help in selecting paints and other organic coatings. Unfortunately, architects, engineers, and building owners often accept a fabricator's standard organic coating without question. Sometimes, for example, the only requirement in architectural specifications is that the finish be the fabricator's standard baked enamel. The results are not always good. Those responsible for selecting or accepting paint or organic coatings should obtain, review, and understand the manufacturer's technical literature for each product to be used, including its contents and application instructions.

Paint that will be applied either in the shop (or factory) or in the field, and other organic coatings that will be applied in the field, must be formulated to dry properly in normal air. Organic coatings that will be applied only in the shop (or factory) may be formulated to be either air drying or as a baking finish. Exterior paint may be either chalking or nonchalking, but it should be mildew resistant.

The quality and expected service life of a paint or other organic coating are related to several characteristics, including color and gloss retention, adhesion, corrosion, mildew, and graffiti resistance, and hiding power (opacity). Another major factor determining quality is the percentage of volume solids in the paint or organic coating material that will remain on the finished surface after the material has dried. The more volume solids there are, the thicker the dried film will be and the greater its opacity. Conversely, the more water or solvent a paint or other organic coating contains, the lower its quality is likely to be. The composition of the solids does, however, have some effect on a paint's or other coating's quality.

The NAAMM says that an optimum organic coating should have good flowing and leveling characteristics, result in a film thickness not less than 1.0 mils (preferably 2.5 mils or thicker) with a high solids content, be fast drying, have a permeable primer (but have high resistance to moisture and gas penetration through the finish coat), exhibit good adhesion characteristics, be flexible, be hard and resistant to abrasion, and be durable. Finding all these traits in one finish is not likely, and some characteristics may be more important in a particular application than others. The finish selected should be the one that has the greatest number of the desired characteristics.

Codes, regulations, or laws often control the acceptable lead content of a paint or other organic coating used in residential applications and other locations where it will be accessible to children.

Paint and Other Organic Coatings' Composition

An old enamel may be nothing more than a pigmented varnish, but modern enamels are practically indistinguishable from paints. There is one difference, however, between paint and other opaque organic coatings as defined in this book's Glossary. This distinction is that the nonpaint opaque organic coatings may have materials added to them, or the basic composition altered, so that forced drying or baking is required for proper setting of the material. With this exception, the modern paint and other opaque organic coatings are composed of four different groups of components. These elements are the vehicle, a volatile or thinner (also called the solvent), the pigment (except in transparent coatings), and various additives. Each group is made up of several different ingredients, each serving its own function.

The Vehicle. The vehicle carries the binder (film former) to the surface to be covered and forms a film to bind the pigment particles together. It also gives the material continuity and makes it adhere to the covered surface.

The type of binder in the vehicle is the characteristic by which most paints and other organic coatings are identified. There are many types of binders used in paints and organic coatings today, the more common being the ones that follow.

Acrylic. Acrylic resins are used to produce high-performance clear coatings for aluminum and copper alloys, the most common of which is probably methacrylate lacquer. Most acrylics used in buildings are air-setting types, though thermosetting acrylics are also available. Incralac, developed by International Copper Research, is an air-setting clear acrylic lacquer often used on copper alloys.

Alkyd. Many enamel paint and other opaque organic coatings used on metal today have an alkyd base. When people today refer to oil paints most really mean one of the alkyds, which are actually oil-modified resins. Alkyd materials are available that will produce either a flat, semigloss, or high-gloss finish. On metals, alkyds produce a strong, long-lasting enamel finish.

Inexpensive clear alkyd coatings are also available. These are used on copper alloys, but they tend to yellow when exposed on the exterior unless they are modified with melamine resins.

Cellulose. There are several cellulose-derived materials used as clear coatings, including cellulose acetate and cellulose ethers. They are used mostly in interior applications, because they tend to darken when exposed on the exterior.

Elastomers. Elastomers are synthetic rubber coatings, such as Neoprene and chlorosulfonated polyethylene. They are highly resistant to chemicals and corrosion.

Epoxy. Two-component epoxy-emulsion paints are used where a high-performance, low-odor paint is required. Modified epoxy resins are also used in high-quality opaque coating systems. They include amine-cured epoxy resins; amine-catalyzed, cold-cured epoxy resins; epoxy-polymide resins; estrified epoxy resins; coal-tar epoxy resins; epoxy–zinc-rich coatings; epoxy esters; and epoxy emulsions.

Clear epoxy coatings are sometimes used on copper alloys in interior applications. They tend to chalk and darken when used on the exterior.

Fluorocarbon Polymers. Coatings made with fluorocarbon polymers, also called just fluoropolymers, have high chemical, abrasion, and wear resistance. They are frequently used today on aluminum. A common fluorocarbon polymer resin is Kynar 500. The better fluorocarbon polymer coatings contain at least 70 percent Kynar 500 resin or its equivalent.

Latex. Paints with a latex film former dry by the evaporation of water. Both interior and exterior paints are available with a latex binder, but other organic coatings do not usually have a latex binder. Latex paints are used on metals in nonresidential buidings, but not as often as are the alkyd types. The latex in latex paints is either polyvinyl acetate, polyacrylic, or polystyrene-butadiene. Latex paints go on easily and dry quickly with little odor. There are many colors available, and their color retention is good.

Nitrocellulose. Air-drying clear nitrocellulose coatings are low in cost and frequently used on copper alloys in interior locations. They are usually modified with acrylic or alkyd resins. In exterior applications they usually require removal and reapplication at approximately one-year intervals.

Oil. Paints with an oil binder were used extensively in the past and are still used today, though less frequently. The drying oil in oil-based paint is usually, but not always, linseed oil. Because they dry slowly, most oil-based paints are now used on exteriors of buildings. They provide high-quality applications over metals.

Oil–Alkyd Combinations. When linseed oil is modified with alkyd resins, a binder results that improves a paint by reducing its drying time, making it harder, and reducing its tendency to fade. Oil-alkyds are used as trim enamels for woods and metals and as primers on structural steel.

Oleoresins. The processing of drying oils with hard resins produces an oleoresinous binder that is used as a varnish or mixing vehicle. These are seldom used today, because they tend to turn yellow over time.

Phenolics. Products containing phenolic binders, which are oleoresinous, are available as pigmented paints or coatings. Coatings are available in dark colors only. Their finish may be either flat or high gloss. Phenolics were among the first synthetic resins. They can be used in wet environments.

Polyesters. Polyesters are generally used in the construction industry to produce what are called "tilelike coatings." They are also sometimes modified to produce metal coatings. Siliconized polyester, for example, is a high-performance coating for metals that some producers tout as being competitive in performance with fluorocarbons (see "Manufacturers and Products" earlier in this chapter).

Other polyester resin coatings are used as industrial coatings on metals.

Rubber-based Binders. While paints with latex binders are water-emulsion types, materials with rubber-based binders are solvent thinned. The resins used in them are actually synthetic rubber. Rubber-based materials are lacquer-type paints that dry quickly to form water- and chemical-resistant surfaces in areas where water is a problem, such as in showers, laundry rooms, and kitchens. They can also be used on exterior surfaces. They are hard to recoat, however, because their strong solvents tend to lift previously applied material.

Rubber-based coatings have characteristics similar to those for rubber-based paints.

Silicone. Clear silicone coatings have high resistance to heat, moisture, acids, salt spray, and alkalis. They are expensive and must be modified for exterior use, to prevent them from darkening.

Urethane. Material with an oil-free urethane binder is used as a pigmented paint or coating. Such urethanes perform the same function as alkyds, but they are more expensive. They are usually used in interior applications, because they often lose their gloss and sometimes become chalky or turn yellow when exposed on the exterior. Clear urethane coatings are used on copper alloys. Older clear urethane coatings were likely to change color, but the newer ones probably will not. Clear urethanes are expensive but highly resistant to abrasion and chemicals.

Vinyl Chloride Copolymers. Coatings made with vinyl chloride copolymers are durable and resistant to moisture, oils, fats, alkalis, and acids.

Combinations. Different types of binders can sometimes be combined to produce binders with characteristics often superior to their components. Some such combinations include phenolic-alkyd, silicone-alkyd, vinyl-alkyd, alkyd-urea-formaldehyde, and alkyd-melamine (alkydamine) binders.

The Volatile or Thinner. Paints and other organic coatings contain low-viscosity volatile solvents or other thinners to make them liquid and help them penetrate the surface to which they are applied. Volatile solvents and other thinners evaporate after the paint or coating has been applied and thus make up no part of the applied film.

Many types of solvents are used in solvent-based paints and other organic coatings. Most alkyd paints and coatings contain alkyd solvents or one or more petroleum-based organic materials such as mineral spirits, naphtha, xylene, or toluene. Other solvent-based paints and coatings contain other solvents such as keytone, acetates, alcohols, or glycol ethers.

In water-reducible paints the thinner (water) separates the binder droplets. In solvent-based paints and coatings the solvent dissolves the binder and holds it in solution. Solvents and water thinners also keep the paint at the proper viscosity and control its setting time. The proper amount of either solvent or water, as appropriate, is essential for proper paint or coating performance. This is why proper thinning of paints by the applicator is essential. It is also the reason that many paint products sold to ordinary consumers, which are called trade sales materials in the industry, require application without thinning.

It is the volatile solvents in paints and other coatings, and their necessary evaporation, that prdouce most of the pollution that paints and coatings contribute to the atmosphere. The solvents used currently are either hydrocarbon or oxygenated types. They both contain hydrogen and carbon atoms, but the oxygenated type also contains oxygen. The hydrocarbon type is cheaper, but its use runs afoul of prevailing laws in some locations and they may eventually give way entirely to the oxygenated type. Both types may ultimately disappear, but replacements for them are not yet available.

Older paints used terpene solvents containing such materials as turpentine or pine oil, but these are seldom used today.

The Pigments. The solid material that is left when a paint or other opaque coating dries is mostly pigment. Pigment particles are permanently insoluble. They dictate the paint or other coating's color, opacity, and gloss, and resist corrosion, weathering, and abrasion of the surface. They also contribute to the paint or other coating's hardness and govern its adhesion to the surface.

A paint or other opaque coating's gloss is determined by its pigment-to-binder ratio. Larger pigment particles produce less gloss. Extender pigments such as clay, calcium carbonate, or silica may also be used to reduce gloss. These extenders and others including iron oxide, talc, and barium sulfate also increase the durability of the paint film. Gloss is rated as flat, eggshell, semigloss, or high gloss. Opaque materials with a sheen are usually called enamel, but flat enamels are also available today.

Opacity is a paint or other coating's ability to cover a surface and hide what lies below the paint. Opacity is a direct function of the pigment used.

Compounds containing lead or zinc as a pigment are common in ferrous metal primers. Such compounds include zinc chromate, zinc yellow, basic lead silico-chromate, red lead, metallic zinc, and zinc oxide. Metallic zinc and zinc dust–zinc oxide paints are particularly effective as primers for galvanized steel. Galvanizing repair paint has a high zinc–metal pigment content, from 85 to 90 percent.

The Additives. Many materials are added to paints and other organic coatings to alter or improve their basic characteristics. Some of these are essential. Among the various additives are plasticizers, driers, oxidizing catalysts, converters, fungicides, preservatives, and wetting, antisetting, and antiskinning agents.

Paint and Other Organic Coating Systems

A single coat of paint or other organic coating on metal will almost never be adequate. Rather, a system of materials must be used in combination to form a satisfactory finish application. All of the coats in a paint or other organic coating system must be compatible for a successful installation. Some paints will permit barrier coats to be applied over incompatible primers, but for most paints and for other organic coatings incompatible primers must be removed and a compatible new one applied.

Most paint and other organic coatings' manufacturers routinely furnish specific recommendations for the several components in their paint or coating system that should be used for most substrates and conditions. The paint and coating systems used should be in accord with those recommendations. In addition, each paint or coating system proposed for use should be reviewed by the manufacturers of the products to be used to ensure compatibility of the various coats of the system with each other and with the materials they will contact, and the suitability of the system for the conditions. When a manufacturer suggests that different products or combinations of products be used other than those selected (a different prime coat, for example), its suggestions should be followed.

Most paint and other organic coating systems consist of at least a primer and a top coat. In many systems one or more intermediate coats are also used.

Unless there is a compelling reason to do otherwise, all the products used in a particular paint or other organic coating system should be produced by the same manufacturer, or at least be supplied by the same company that manufactures the finish coats. Thinners should be approved by the manufacturer of the material being thinned. Undercoats and primers should be approved by the finish-coat manufacturer, whether they will be shop- or field-applied.

Primers are paints or other organic coating system coats that have been formulated for application to bare metal as a base for succeeding coats. They must be formulated to adhere readily to the metal and the next coat. They must be compatible with both the metal and the top coats. Oil-based and alkyd primers should not be used on galvanized surfaces, for example, because their presence can create a chemical reaction that leaves a soap on the metal's surface that will not permit paint to adhere. Some primers, such as the rust-inhibitive ones, also serve the major secondary function of protecting the metal. Some primers also isolate essentially incompatible paint and substrates. For example, primers on galvanized metal surfaces separate the galvanizing from the top coats, many of which contain chemicals that can react with the zinc coating on the metal.

Ferrous metals that will be painted in the field are usually shop primed, unless they are galvanized or aluminized, but even these metals are sometimes shop primed with paint. Galvanized surfaces to which a shop coat of paint will be applied should first be given a phosphate or equivalent chemical treatment, to provide a bond for the paint.

There are several types of paint that have enjoyed great success as primers on raw ferric metals. They include but are not necessarily limited to the following: red lead mixed pigment, alkyd varnish, and linseed oil paint conforming with Federal Specification TT-P-86G, Type II; red lead iron oxide, raw linseed oil, and alkyd paint conforming with the Steel Structures Painting Council's SSPC Paint 2-64; and basic lead silico-chromate-base iron oxide, linseed oil, and alkyd paint conforming with Federal Specification TT-P-615, Type II. These materials are usually applied at a rate necessary to form a final primer thickness of about 2.0 mils. Two coats are usually applied to surfaces that will be in contact with exterior masonry or concrete.

Because they contain lead, some of the above-listed paints may not be acceptable today in some jurisdictions for some uses.

Primers for galvanized metals are usually metallic zinc paints or zinc dust–zinc oxide paints. The federal specification for this type of primer is TT-P-641G.

Whatever the primer used, it must be compatible with both the substrate metal and the finish coats of paint. The selection of a metal primer must be coordinated with the requirements of the finish paint's manufacturer.

Top coats are the main barrier to weather, chemicals, soiling, and abuse. They also protect primers and provide a finished surface.

Special Paints

Paints specifically formulated to handle unusual conditions are also available. For example, the paints intended for use on cleaned metal are generally called direct-to-metal paints. There are also direct-to-rust paints, which are epoxy mastic materials formulated for direct application over rust. There are also rust-conversion coatings, which convert ferric oxide (rust) to a stable organic iron compound that then becomes part of the coating.

Other special paints include aluminum, bituminous, emulsified asphalt, and chlorinated-rubber paints. These materials are used on metal exposed to particularly harsh conditions.

The bituminous paints, which may contain coal-tar pitch, asphalt, linseed oil, and oleoresinous varnishes, are usually required to comply with the requirements of the Steel Structures Painting Council's SSPC Paint 12 (cold-applied asphalt mastic). Bituminous paints are often used as corrosion protection for steel, aluminum, and copper alloys to be in contact with a material that might cause them to corrode, such as concrete, masonry mortar, or plaster. Bituminous paints are not attractive and are, therefore, usually used where they will be concealed in the completed work.

Powder Coatings

Powder coatings are applied by covering a metal surface with a fine powder, then fusing the powder into a continuous film coating by heating it to temperatures between 300 and 400 degrees Fahrenheit. Most powder coatings used today are based on epoxy, polyester, or acrylic resins. Powder coatings are relatively new in this country and will probably not be found on buildings constructed before 1982.

Film Laminates

Both interior and exterior finishes on copper alloys are sometimes protected by a laminated film called Incracoat. This film, developed by the Copper Research Association, consists of a 1-mil-thick layer of polyvinyl fluoride, which is bonded to the copper alloy with an adhesive.

Strippable Coatings

Temporary clear coatings are used on some metals, particularly aluminum, to protect the finish during handling and installation. These coatings must usually be removed soon after the metal has been installed, especially in exterior applications. Heat and sunlight will make some of them difficult to remove if they are left exposed.

Color, Gloss, and Texture

For a given job a detailed color schedule identifying each color to be used should be made up and provided to the applicator. Color pigments should be pure, nonfading, applicable types to suit the substrates and service required.

Each coat in a paint or other opaque organic coating system should be a slightly different shade from the preceding coat. The final colors, glosses, and textures should match acceptable samples.

Organic Coating Application

Preparation for Paint and Other Organic Coating Application

Regardless of the type of organic finish to be applied, the metal must be prepared to receive it, and the conditions under which it is applied and cured must be in accordance with the manufacturer's recommendations.

Inspection. Materials to be painted or to receive another organic coating, and the conditions under which the paint or organic coating will be applied, should be inspected carefully. Unsatisfactory conditions should be corrected before work related to paint or other organic coatings is begun.

Each coat of a paint or coating system should be inspected and found satisfactory before the next coat is applied.

Preparation of Metal Surfaces.   Surface preparation is certainly among the most important aspects of organic coatings application. It will often determine whether an application will be successful (Fig. 5-1). Some sources indicate that as high as 80 percent of paint failures are caused by improper surface preparation.

Hot-rolled ferrous metal comes from the mill with a coating of mill scale and rust. A few types of paint may be applied by brush directly over this scale and rust, but the vast majority of organic coatings require complete removal of this residue to be successful. Cold-rolled surfaces must be degreased before finishing and may also need roughening if they are too smooth. Aluminum ordinarily leaves the mill with a coating of oil and grease, and the process of making nonstructural metal items leaves weld deposits and other contaminants on their metal surfaces. All such foreign materials must be removed or properly treated before an organic coating is applied.

Figure 5-1   The paint on this railing is flaking off because the original paint was not properly prepared.

Metal surfaces to receive paint or another organic coating must be cleaned of oil, grease, dirt, loose mill scale, and other foreign substances, using solvents or mechanical cleaning methods. The recommendations of the paint or coating's manufacturer and of the Steel Structures Painting Council should be followed. Bare and sandblasted or pickled ferrous metal should be given a metal-treatment wash before primer is applied. Refer to the discussion of each particular metal in Chapter 4 for that material's pretreatment and cleaning requirements before applying an organic coating.

Preparing a Site for Field Application of Paint or Other Organic Coatings. Hardware, accessories, machined surfaces, plates, fixtures, and similar items that are not to be painted or coated should be removed before surface preparation begins. Alternatively, such items may be protected by taping over their surfaces or by another type of protection. Even when such items are to be painted or coated they are often removed, to make painting or coating them and adjacent surfaces easier. After the painting or other coating application of each area in question has been completed, the removed items can be reinstalled.

Surfaces should be cleaned before the first coat of paint or other organic coating is applied and if necessary between coats to remove oil, grease, dirt, dust, and other contaminants. Activities should be scheduled so that contaminants will not fall onto wet newly painted or coated surfaces. Paint or other organic coatings should not be applied over dirt, rust, scale, oil, grease, moisture, scuffed surfaces, or other conditions detrimental to the formation of a durable coating film.

Mildew should be removed and affected surfaces neutralized. Some references recommend scrubbing mildewed surfaces thoroughly with a solution made by adding two ounces of a trisodium phosphate-type cleaner and eight ounces of sodium hypochlorite (Clorox) to one gallon of warm water. A scouring powder may be used if necessary to remove mildew spores, but care must be taken to prevent damage to surfaces being cleaned. Cleaned surfaces should be rinsed with clear water and allowed to dry thoroughly. Some jurisdictions, however, do not permit the use of these chemicals. In those locations it will be necessary to ascertain and use legal materials. Contact the coating's manufacturer for advice.

Before painting or coating is started in an area, that space should be swept clean with brooms and excessive dust removed. After an application of paint or other organic coating has been started in a given area, that area should not be swept with brooms. Necessary cleaning should be done from this point on using commercial vacuum-cleaning equipment.

Surfaces to be painted or coated should be kept clean, dry, smooth, and free from dust and foreign matter that would adversely affect the paint or other coating's adhesion or appearance.

Applying Paint and Other Organic Coatings

General Application Requirements for Organic Coatings. The type of material chosen for each coat, location, and use should be that designated for the purpose by the paint or other organic coating material's manufacturer in its published recommendations.

Paint and other organic coating materials should arrive at the application site ready for application, except for tinting and thinning. They should be in their original unopened containers bearing the material's name or title, manufacturer's name and label, standard to which it complies (federal specification number, for example), manufacturer's stock number, date of manufacture, contents by volume for its major pigment and vehicle constituents, thinning instructions, application instructions, color name and number, and fire-hazard data, where applicable.

Painting and other organic coating materials should be mixed, thinned, and applied in accordance with the manufacturer's latest published directions. Materials should not be thinned unless the manufacturer specifically recommends doing so. Applicators and techniques should be those best suited for the metal and the type of material being applied.

Materials not in actual use should be stored in tightly covered containers. Containers used for the storage, mixing, and application of paint and other organic coatings should be maintained in a clean condition, free of foreign materials and residue.

Unless the manufacturer recommends otherwise, paint and other organic coating materials should be stirred before application, to produce a mixture that is of uniform density, and be stirred as required during application. Surface films should not be stirred into the material. Such film should be removed and, if necessary, the material be strained before use. Shaking to mix some materials is permitted by some manufacturers, but shaking some types of organic coatings will introduce air bubbles that will remain in the applied coating and spoil the finish. The manufacturer's advice should be strictly followed with regard to the shaking of organic coatings to mix them.

Workmanship should be of a high standard, and application should be done by skilled mechanics. Field applications should be done using proper types and sizes of brushes, roller covers, and spray equipment. Shop applications should be made using the appropriate methods.

The proper equipment should be used to apply paint and other organic coatings, and it should be kept clean and in proper condition. For example, rollers or brushes used to paint concrete or masonry should not be used to paint metal. Rollers used for a field-applied gloss finish and its corresponding primer should have a short nap.

Material should be applied evenly and uniformly, under adequate illumination. Surfaces should be completely covered and be smooth and free from runs, sags, holidays, clogging, and excessive flooding. Completed surfaces should be free of brush marks, bubbles, dust, excessive roller stipple, and other imperfections. Where spraying is either required or permitted, the paint or other organic coating should be free of streaking, lapping, and pileup.

The number of coats recommended by the manufacturer for each particular combination of use, substrate, and paint or other organic coating system should be the minimum number used. When properly applied, that number of coats should produce a fully covered, workmanlike, presentable job. Each coat should be applied in heavy body, without improper thinning.

Paint primer coats may be omitted from previously primed surfaces, but every other coat recommended should be applied to every surface to be covered. When stains, dirt, or undercoats show through the final coat of paint or other opaque coating, the defects should be corrected and the surface covered wih additional coats until the paint or coating film presents a uniform finish, color, appearance, and coverage. When stains, dirt, and other imperfections show through a clear coating, the coating and the offending imperfection should be removed and the cleaned area recoated.

Paints and other organic coatings should be applied at rates of coverage sufficient for the fully dried film thickness of each coat to be not less than that recommended by the manufacturer. Special attention should be given to ensure that edges, corners, crevices, welds, and exposed fasteners receive a dry-film thickness equivalent to that on the flat surfaces.

The first coat of paint or other organic coating should be applied to surfaces that have been cleaned, pretreated, or otherwise prepared for finishing as soon as it is practicable to do so after surface preparation and before it can begin to deteriorate.

At the completion of the application process, paint and other organic coatings should be examined. Damage found should be touched up and restored so that the painted or coated surface is left in an acceptable condition.

Shop Primer-Coat Application for Field-Applied Paint. Ordinarily, a shop coat of paint is applied to all ungalvanized, nonstructural ferrous-metal items that are not to receive a complete finish in the shop. Shop coats might also be applied to aluminum or copper-alloy items when they are to be field painted. Most aluminum items that receive an organic coating have all the coats applied in the shop. Copper alloys are, however, seldom finished with an applied organic coating. Galvanized and aluminized steel are not usually given a shop coat but may be. Members and parts of members that will be embedded in concrete, stone, or masonry are not usually given an organic coating, except for coatings used strictly to prevent metal corrosion. Refer to "Types of Corrosion" in Chapter 3 for a discussion on the use of coatings to prevent corrosion. Surfaces and edges that are to be field welded should not be given a shop coat of paint or other organic coating.

A shop coat of primer paint should be applied only to cleaned, degreased metal surfaces. As discussed earlier in this chapter and in Chapter 4, immediately after surface preparation the primer should be brushed or sprayed on, in accordance with the manufacturer's instructions, at a rate that will produce a uniform dry-film thickness of about 2.0 mils for each coat. The painting methods used should give full coverage of joints, corners, edges, and exposed surfaces.

Only one shop coat is usually applied on fabricated metal items, but two coats are normally applied on surfaces that will be inaccessible after assembly or erection of the metal item. It is a good idea always to change the color of the second coat so that an observer can distinguish it from the first, and thereby determine if all required coats have been completely applied.

Field Touch-up of Shop Coats.   The first step in preventing metal-finish failure from corrosion is to make sure that the factory or field finish originally applied is touched up to prevent corrosion from occurring at scratches and abrasions. The initial touch-up should be repeated whenever needed throughout the life of the finish.

Welds, abrasions, bolted connections, factory- or shop-applied prime coats, and shop-painted or -coated surfaces should be touched up after erection and before calking and application of the first field coat of paint or other organic coating. The touch-up material should be the same as the primer. Surfaces that are to be covered should be touched up before they are concealed. Touch-ups in paint should be sanded smooth.

Before touch-up is started, surfaces should be cleaned. Contaminants such as mortar, damaged or unwanted paint, dirt, grease, dust, effloresence, salts, and stains should be thoroughly removed. The cleaning materials used should not harm the metal, metal finishes, or adjacent surfaces. Abrasives or caustic or acid cleaning agents should not be used. Initial cleaning should be done using only clear water and mild soap.

Immediately after a metal item has been installed, the shop paint should be removed completely from field welds, bolted connections, and abraded areas. The exposed areas should be painted with the same material used for shop painting. The new paint should be applied with a brush or spray to provide a minimum dry-film thickness of 2.0 mils.

Application of Paint.   In addition to the general requirements stated thus far, paint should be applied in accordance with the recommendations of the paint manufacturer and the points that follow.

Paint for field painting should be stored in a well-ventilated single place for each major portion of the work. These storage areas should be kept clean. Paint materials should not be stored in closets or other small, confined locations.

Oily rags, oil, and solvent-soaked waste should be removed from buildings at the end of each work day. Precautions should be taken to avoid the danger of fire.

Water-based paints should be applied only when the temperature of the surfaces to be painted and the surrounding air is between 50 and 90 degrees Fahrenheit, unless the paint manufacturer's printed instructions say otherwise.

Solvent-thinned paints should be applied only when the temperature of the surfaces to be painted and the surrounding air is between 45 and 95 degrees Fahrenheit (some sources say 50 to 120 degrees), unless the paint manufacturer's printed instructions recommend otherwise.

Paint should not be applied in snow, rain, fog, or mist or when the relative humidity exceeds 85 (some sources say 90) percent; or to damp or wet surfaces; or to extremely hot or cold metals unless the material manufacturer's printed instructions recommend otherwise. The painting of surfaces exposed to the hot sun should be avoided, but painting may continue during inclement weather if the areas and surfaces to be painted are enclosed and heated to within the temperature limits specified by the paint manufacturer during the application and drying periods.

Once painting has been started within a building, a temperature of 65 degrees Fahrenheit or higher should be provided in the area where the work is being done. Wide variations in temperatures that might result in condensation on freshly painted surfaces should be avoided.

It is necessary to sand lightly between each succeeding coat of most paints used on metals, especially paints that have a gloss finish. High-gloss paint should be sanded between coats with very fine grit sandpaper. Dust should be removed after each sanding, to produce a smooth, even finish.

Bronze and aluminum surfaces should be protected from corrosion, by coating the contact surfaces with bituminous paint, where they will be in contact with dissimilar metals, concrete, stone, masonry, or pressure-treated wood. Alternatively, such metals may be coated with a zinc-chromate primer paint.

Weatherstripped doors and frames should be finished before door equipment is installed.

The first coat (primer) recommended by the paint manufacturer may be omitted on metal surfaces that have been shop primed, provided that the primer is touched up in the field to repair abrasions and other damage.

Sufficient time should be allowed between successive coatings to permit proper drying. A minimum of twenty-four hours is required between interior coats, forty-eight hours between exterior coats. Surfaces should not be recoated unless the previous coat has dried until it feels firm, does not deform or feel sticky under moderate thumb pressure, and the application of another coat of paint will not cause lifting or loss of adhesion of the undercoat.

The edges of paint adjoining other materials or colors should be made sharp and clean, in straight lines and without overlapping.

Mechanical and electrical equipment and devices that have not been factory finished should be painted as follows. Pumps, fans, and heating and ventilating units should receive two coats of paint. The interior surfaces of ducts and chases where visible through registers or grilles should be painted with a flat, nonspecular black paint. Exposed brass or copper pipe should be painted the same as if it were steel and iron. Nameplates, moving parts, and polished surfaces should not be painted. Hangers and supports should also be painted while pipes and equipment are being painted.

Paint should be cured in the proper humidity and temperature conditions, as recommended by the materials' manufacturers.

Application of Clear Organic Coatings.   Many of the requirements just given for paint application apply as well to the field application of clear organic coatings. Seldom, however, are clear organic coatings applied to new metals in the field, because it is usually more economical and easier to apply coatings on new materials in the shop.

Shop applications should be made in accordance with the recommendations of the finishing material and metal manufacturers, and the applicable industry standards. There is one notable exception, however. Regardless of who recommends that it be done, methacrylate or other lacquer should never be used on color-anodized aluminum. If it is, the lacquer will tend to alter the color and make the anodizing appear more like paint than anodizing. In addition, lacquer never completely adheres to anodized aluminum. Where it does not fully adhere it will appear white. On clear anodized aluminum the difference is not noticeable, but on color-anodized aluminum the white spots produce a blotchy, unappealing effect. If protection of color-anodized aluminum is necessary, a strippable coating should be used. Strippable coatings are expensive to apply and remove, however. If they are left on too long in the sun, they may become almost impossible to remove. Covering the aluminum with a loose plastic film, paper, or even grease will usually offer ample protection.

Before a sealant is applied, any protective strippable coating or tape on aluminum or on copper alloys should be removed from the portion of the metal that will be in contact with the sealant.

Although the use of clear coatings is questionable on aluminum, they are frequently used on copper-alloy surfaces. They should be applied strictly in accordance with their manufacturers' instructions and the recommendations of the CDA.

Other Organic Coating Applications. In addition to the general requirements stated earlier in this section, under ''General Application Requirements for Organic Coatings,'' other organic coatings should be applied strictly in accordance with the recommendations of the manufacturers of the coating product and the metal, and the fabricator's standard practices.

Other organic coatings are usually applied by such mechanical methods as hot spraying, airless spraying, electrostatic spraying, dipping, flow coating, or roller coating. The latter is a coil-coating system used often to finish sheet steel and aluminum used in manufacturing building panels and similar items. Refer to ''Where to Get More Information'' at the end of this chapter for sources of data about coil coating.

Baking-finish organic coatings dry either by evaporation of their solvents in the same way most paints dry or by chemical reaction. In any case, increased temperatures are used. When the temperatures are below 200 degrees Fahrenheit, the process is called forced drying. When the drying takes place above 200 degrees Fahrenheit, the process is called baking. Most baking is done at temperatures between 250 and 600 degrees Fahrenheit, depending on the coating.

Why Organic Coatings on Metals Fail

Because nonstructural metals in buildings are always supported by or attached to something, some failures of organic coatings applied to those metals can be traced to a problem with the supporting or underlying materials or systems. The possibilities include one or more of the following: concrete or steel structure failure, concrete or steel structure movement, wood framing or furring problems, solid substrate problems, and other building element problems, as discussed in Chapter 2 under these headings, which are listed in the Contents. Many of these problems may not be the most probable cause of the failure of organic coatings on metal but may be more serious and costly to fix than the types of problems discussed in this chapter. Consequently, the possibility that they may be responsible for a metal coating's failure should be investigated.

The causes discussed in Chapter 2 should be ruled out or repaired, if found to be at fault, before metal coating repairs are attempted. It will do no good to repair an organic coating on a metal surface or to recoat the metal of a failed, uncorrected, or unaccounted-for problem of the type discussed in Chapter 2 exists, because the repaired or recoated coating will also fail.

Other possible causes of finish failure are discussed in Chapter 3 under "Errors That Lead to Corrosion" and "Errors That Lead to Fracture," and the text that precedes them. Those causes should be investigated and ruled out before repair or refinishing of the failed coating is attempted.

Still more possible causes of coating failure are addressed in Chapters 6 and 7, under a number of different headings. Refer to "Evidence of Failure in Organic Coatings on Metal" later in this chapter for the applicable headings in Chapters 6 and 7.

After the problems discussed in Chapters 2, 3, 6, and 7 have been investigated and found not to be present or have been repaired if found, the next step is to discover any additional causes for the organic coating failure and correct them. Some possible additional causes include bad materials, immediate substrate failure, improper design, bad workmanship, failure to protect the installation, poor maintenance procedures, and natural aging. The following sections discuss these additional causes. Refer to "Evidence of Failure in Organic Coatings on Metal" later in this chapter for a listing of the types of failure applicable to the causes addressed below.

Bad Materials

There are several ways in which the materials used in a failed coating may have been bad. The underlying metal itself may have been improperly produced. The paint or other organic coating materials may also have been improperly manufactured. Improperly produced metals or paints and other organic coating products are certainly not unheard of, a possibility that should be considered when an organic coating on a metal fails. The number of incidents of bad materials is small, however, compared with the number of cases of bad design and workmanship. The various types of manufacturing defects that might occur include those discussed in the following paragraphs.

The metal materials may have been improperly produced. Steel, for example, may have the wrong carbon content. Metal may have incorrect amounts of the various constituents in an alloy. The material may have been improperly quenched, annealed, or heat treated. There are many possibilities, but bad metal itself is not a frequent cause of organic coating failure.

Among other causes are that there may have been too much oil in a paint or other organic coating, which can cause alligatoring and checking.

The basic paint or other organic coating material may also be inconsistent in color, composition, or density.

The pigments in a paint or other opaque organic coating may be incompatible with its other ingredients.

A paint or other organic coating may contain too much pigment for the binder. Such materials will chalk excessively.

The formula of a paint or other organic coating may be such that the material does not fully dry, or at least dries very slowly. Using old or defective materials can have the same effect, as can not having enough drier in a paint or using a poor quality solvent that vaporizes too slowly. A slow-drying oil will also cause paint to dry too slowly. Poorly formulated paint may never dry. An organic coating with a poorly formulated drying finish may not cure at the usual baking temperature and may, therefore, not be properly cured when delivered to the construction site.

Immediate Substrate Failure

Failure in the underlying metal or in an underlying mechanical or chemical finish will utlimately result in the failure of an applied organic coating. If, for example, the substrate material should crack, applied paint or other organic coatings will also crack. The types of damage to the underlying metal or finish that may lead to finish failure include those that follow.

Corrosion damage to an underlying material is one such problem (Fig. 5-2). When a fracture is involved, it is often difficult to determine whether a fracture let the corrosion begin or the corrosion caused a fracture to form. The two are usually linked in some way. Refer to ''Errors That Lead to Corrosion'' and ''Errors That Lead to Fracture'' in Chapter 3, and the pages preceding them, for a detailed discussion. It follows that if steel rusts beneath a paint or coating the paint or coating will fail, whether or not there is additional damage (fracture) to the steel caused by the corrosion (Fig. 5-3).

Fracture of the underlying metal may result from many factors unrelated to corrosion. Regardless of its cause, metal fracture will always lead to the failure of an applied organic coating. Refer to ''Errors That Lead to Fracture'' in Chapter 3 and the pages that precede it for a discussion of fracture. Chapters 6 and 7 also discuss possible reasons for fracture.

Delamination or other failure of an underlying mechanical or chemical finish will usually cause an applied organic coating to fail. This will be a problem regardless of the type of underlying finish, but it is more likely to happen when an underlying metallic coating fails. If, for example, the aluminum coating in an aluminized steel product delaminates from the steel, any secondary applied finish will also fail. Chapter 4 discusses the types of failure that may occur in finishes that may be used beneath an organic coating.

Figure 5-2 The metal at the base of the fence post and brace have deteriorated underneath the paint. The paint is just beginning to show damage.

Improper Design

Improper design includes the types of errors discussed in Chapter 6 under the same heading, but those errors concern the design of the nonstructural metal item in which the metal materials discussed in Chapter 3 are used with the finishes discussed in Chapters 4 and 5. Improper design also includes selecting the wrong paint or other organic coating material for the location and conditions, as well as requiring improper installation of those materials. The latter type of design problems includes those now discussed. Any one of them may lead to the failure of an organic finish on metal.

Selecting a paint or other organic coating material having a composition inappropriate for the location and use intended will usually lead to failure. Using residential quality paint in a commercial lobby is one example. Se-

Figure 5-3   The metal beneath this finish has rusted. It is often difficult to tell whether the finish failed, letting the rust begin, or the rust started where a finish had not been applied, as on the concealed side of the panel. In this case the finish failed first.

lecting latex paint for use where the humidity will be high is another. High humidity can cause the water-soluble components in latex paints to appear as brown spots in the dried paint.

Selecting incompatible paint-system products will also lead to failure. Using a water-based top coat over a solvent-based undercoat, for example, may cause alligatoring, checking, or peeling. Using a hard finish over a soft undercoat can result in alligatoring or checking. Latex paint applied over old, chalking oil paint cannot penetrate the chalk and thus probably will not adhere. Applying an oil paint over a latex can cause separation, because when they age the oil paint becomes harder and less elastic than the latex. Incompatible top coats may also blister.

Selecting paints or other organic coating materials that are incompatible with their substrates, including existing finishes, may lead to peeling, flaking, cracking, or scaling of the coating.

Selecting paints or other organic coating materials of inferior quality may lead to failure. An example is selecting a paint that has excessive chalking characteristics. Another is selecting a chalking paint for use where runoff will cross another material, such as brick.

Failing to require the removal of paint that is more than 1/16 inch thick can lead to failure. Thick undercoats can cause new top coats to alligator or crack.

Failing to require proper preparation or application methods can cause every type of coating failure possible. Most preparation methods are addressed in Chapter 4.

Failing to require that a proper number of coats be applied can produce a coating through which the substrate is visible, or which does not protect the coated metal, or both.

Failing to require the proper treatment and protection of metallic-coated steel will usually result in an applied organic coating that peels, flakes, or otherwise delaminates (Fig. 5-4). When the metallic-coated metal is not finished with an organic coating, the failure to treat and protect it properly can lead to corrosion of the underlying metal. There is a tendency, for example, to believe that galvanized steel is immune from corrosion, which is far from the truth. The zinc coating on galvanized metals is itself subject to white rust, which will interfere with paint adhesion. The zinc coating is also subject to galvanic corrosion from more noble metals, such as copper, and chemical corrosion from soluble sulfites like those found in cinders. Zinc will also be corroded by salt, some concrete aggregates, and the acids in cedar and oak. The zinc coating itself may oxidize or fail to protect its ferrous metal core when it is submerged in water or subjected to heavy layers of condensation or other water deposits. Zinc coatings will be burned away during welding. The preparation of metallic-coated metals is addressed in Chapter 4.

Bad Workmanship

Bad workmanship in fabricating and installing nonstructural metal items is discussed in Chapters 6 and 7, respectively. Correct preparation for and installation of paint and other organic coatings is also essential to prevent their failure. This section discusses workmanship problems that are directly related to paint and other organic coating applications and that can adversely affect them.

Probably the biggest cause of failure from bad workmanship is failure to follow the design and recommendations of the manufacturer and the relevant recognized authorities. For example, applying fewer than the recommended number of coats can result in failure to cover the surface and in the telegraphing of underlying faults through the paint.

It is also asking for failure in a paint to apply it or another air-drying

Figure 5-4 Improper pretreatment of this galvanized steel louver almost certainly caused its paint to peel and flake where the bright spots show.

organic coating in a building before the building has been completely closed in and before wet work, such as concrete, masonry, and plaster, has dried out sufficiently. Moving a factory-coated metal into the same type of environment may also have detrimental effects on its coating. Such conditions may cause the humidity to be too high or form condensation on surfaces, whether they are still to be coated or are already coated. Applying most paints and other organic coatings to damp or wet surfaces can cause them to adhere poorly and blister. Also, applying them when the humidity exceeds the level recommended by their manufacturer can cause them to dry slowly and not adhere or form blisters. In latex paints, brown spots may appear when the humidity is too high. Properly ventilating spaces where paint or other organic coatings are being applied not only protects workers from an accumulation of potentially toxic fumes but also helps control humidity and promotes the proper drying of paint and other organic coatings. Placing factory-coated metal items in environments for which their finishes were not designed will almost always lead to coating failure.

Installing paint or other organic coatings with a room temperature below that recommended by the coating material's manufacturer for a minimum of forty-eight hours before installation, or when the paint or other organic coating material's temperature is too low, can cause failure. The manufacturer's recommended minimum temperature for field-applied, air-drying paints and other organic coatings generally ranges from 45 to 65 degrees Fahrenheit and will also vary with paint or coating type.

An associated problem is that of letting the room temperature fall below the temperature recommended by the manufacturer after the paint or other coating has been applied. Low temperature is a major factor in air-drying organic coating failures, because it can affect drying time and adhesion. Low temperatures shorten drying time and make a paint or other organic coating hard to apply. Applying a paint or other organic coating containing ice or applying them to a frozen or frost-covered surface can lead to poor adhesion and blistering. Low temperatures can cause wrinkling and prevent a coating from adhering.

Another major error producing failure is to apply paint or another field-applied organic coating when the temperatures of the air, the surface being coated, or the coating material are too high. High temperatures can thin a coating material and make it not cover as well as it would otherwise. Air-drying organic coating materials that are too warm when applied will set too rapidly and form too thin a film. One source of excess heat is hot air passing over the surface being coated. This can happen, for example, when a temporary heater is placed too close to the surface. Such a passage of hot air not only raises the surface's temperature but can also cause the surface to dry too rapidly through evaporation, which traps wet coating material beneath a dried surface. Later, when the underlying coating material does dry, it will shrink and sometimes delaminate from the top coat. Another potential result is the trapping of vapors that evaporate from the undercoat, either between the two coats or between the lower coat and the substrate. In either case bubbles form, which may rupture and form pits or puncture wounds that can extend completely through the paint to the substrate. Applying paint or other organic coatings in direct sunlight or when the air temperature is too high may also cause the surface of the coating to dry too quickly, entrapping solvent vapors that will appear as blisters (Fig. 5-5). Coating materials, surfaces, or air that are too hot can also cause a coating's surface to wrinkle or the coating not to adhere to the substrate.

The manufacturer's recommended maximum temperatures for the application of paint and other organic coatings usually range from 85 to 120 degrees Fahrenheit, but they may vary with the paint type.

Figure 5-5 The blisters and flaking paint on this railing were almost certainly caused by painting the metal when it was too hot.

Another cause of failure is permitting drastic changes to occur in temperature where a paint or other organic coating has been applied, or in an area into which a factory-coated metal item has been placed. When drastic temperature changes occur before a paint or other air-drying organic coating has completely set, alligatoring or checking may occur as the paint expands or contracts. If the top coat is not elastic enough it will crack.

Another temperature-related error is that of permitting a wide temperature differential to exist between the paint or other organic coating and the surface to which it is applied. Large differences may cause the paint or other coating material to blister and eventually peel.

Failing to properly prepare an area where a paint or other organic coating material will be installed can also contribute to failure. Most preparation requirements have been addressed in Chapter 4. It is necessary to remove mildew, moss, ivy and other plant growth, oil, grease, dirt, dust, rust and other corrosion, loose mill scale, wax, loose existing paint, stains, and other contaminants that will interfere with proper application, damage the paint or other organic coating, or telegraph through the paint or other organic coating material. Coatings applied over grease or oil may dry slowly—or not at all. Because many paints and coatings permit water vapor transmission, improperly prepared metals that begin to corrode may continue to do so beneath the coating (Fig. 5-6) and eventually disappear almost completely, leaving a hollow shell of coating material. Even in less severe cases, corrosion beneath a coating may stain the coating to create unsightly surface irregularities (Fig. 5-7).

It is also necessary to smooth out rough surfaces and roughen surfaces that are so smooth that paint or another organic coating will not adhere, and to ensure that surfaces are dry and that no other condition exists that is detrimental to the formation of a durable paint or finish film.

Failing to remove sources of moisture and water that will affect the paint or other organic coating, and to ensure that the metal is completely dry, may cause a paint or other organic coating to dry slowly. Not doing so will aid mildew, moss, and other plant growth and cause the paint or other organic coating to blister and eventually peel, flake, crack, or scale.

Figure 5-6 The metal is corroding beneath the paint coating shown in this photo.

Figure 5-7 The corrosion beneath the paint in this photo has created an unsightly appearance.

In clear organic coatings, especially shellac, alligatoring is often caused by exposure to damp conditions.

The failure to remove dust, dirt, moisture, and other contaminants between coats of paint or other organic coatings can cause the paint or other organic coating to become loose, peel, crack, flake, or scale. Moisture left between coats can cause a paint or other organic coating not to dry properly and may cause blisters that eventually lead to interlayer peeling.

Failing to remove natural salts from previously painted surfaces will result in peeling of an applied paint or other organic coating. The salts that normally collect on exterior surfaces are washed off by rain from exposed surfaces. They must be removed, however, from areas that rain does not touch. Most sources recommend washing such surfaces with trisodium phosphate or another material recommended by the organic coating's manufacturer. Some jurisdictions may not permit the use of certain chemicals, however, so it is necessary also to check legal restrictions before selecting the chemicals to use.

Failing to remove loose, peeling, or otherwise unsound paint or other organic coating material from an existing surface before applying new paint or organic coating materials will also result in failure of the new coating.

If the gloss is not removed from an existing surface or undercoat before a succeeding coat of paint or other organic coating is applied, the paint or other organic coating will not adhere. Applying new paint over existing glossy paint without roughening the surface is one example. Others are applying paint directly over a vitreous coating, slick laminated coatings, stainless steel with a highly polished finish, or chromium plating.

Failing to sand previous coats of paint before the next coat is applied and remove the dust generated can result in lack of adhesion between coats.

Failing to remove chalking before repainting will result in failure of the new paint to adhere to the old.

Failing to remove rust or other corrosion, and to apply a rust-inhibitive primer or one of several special paints formulated for application over rust, can result in continuing corrosion.

Not sanding shoulders at the edges of sound paint or feathering the edges of existing paint where new paint is to be applied will result in the old edges telegraphing through the new paint.

Using paint or other organic coating materials that are of the wrong type for the installation at hand will usually lead to failure. The type of failure will depend on the materials used and how severely they are mismatched.

A similar problem occurs when a coat of paint or other organic coating is applied that is incompatible either with previous coats, an existing finish, or the substrates.

A slightly different but related problem is that of using the wrong primer for a substrate or finish coats, which can cause many types of problems. For example, some types of paint will react with the zinc on galvanized metal to create a type of film that will prevent paint from adhering.

Applying damaged paint or other organic coating materials will usually lead to failure of the coating, regardless of whether the damage was inherent in the manufactured materials or occurred during shipment, storage, or application. Applying paint or other organic coatings that are stained, for example, will result in a discolored surface.

The same applies to the individual components of a paint or other organic coating. For example, using the wrong thinner or a defective or poor quality thinner will cause a coating to fail. Such solvents may evaporate too quickly, resulting in wrinkling, alligatoring, blistering, peeling, flaking, cracking, or checking of the coating. Or they may dry slowly, resulting in a paint or other air-drying organic coating that dries too slowly or never completely.

Using too little thinner in a paint can result in peeling, flaking, cracking, or scaling. Conversely, using too much thinner may result in uneven colors. This is more of a problem with alkyd- or oil-based paints than with water-based materials. Adding too much thinning material in order to save paint may result in premature aging, fading, or an inability of the paint to withstand normal cleaning.

Failing to agitate and mix paint or other applicable organic coatings properly before and during application can cause color separation. Insufficient mixing may result in pigments not being properly blended, which in turn may cause alligatoring, checking, peeling, flaking, cracking, or scaling. Improperly mixed paints and other organic coatings may not dry.

Mixing incompatible materials together will usually result in material failure. How they fail will depend on the materials that are mixed together.

Adding lampblack or another shading material to professional paint materials to make a paint hide better with less paint can result in the paint's failing to adhere properly, changing in color, or prematurely aging.

Adding too much oil to paint can result in alligatoring or checking and may cause the paint to dry improperly. This situation can occur inadvertently if spraying equipment is used when there is grease or oil in the spray line.

Using pigments that are incompatible with other ingredients can result in alligatoring, checking, peeling, flaking, cracking, or scaling.

Failing to apply paint or other organic coating materials properly, including not using enough material, improperly applying it, and not applying the correct number of coats, can lead to failure. Typical examples include failing to cover a surface completely, and applying coats that are too thin or too thick. Applying too-thick coats will cause the surface film to form before the paint beneath it has dried. At first the film will be smooth across the surface, but when the underlying paint dries it shrinks, which will wrinkle the surface film. Paint and other organic coatings' coats that are too thick may also check or even alligator. Too thick a paint coat, especially in an undercoat, can also cause peeling, flaking, cracking, or scaling. Paint or other organic coating materials that are applied in too-thick coats may also run or sag and not dry properly.

Applying a coat of paint or other air-drying organic coating to a still-wet undercoat can cause the surface paint to dry slowly or wrinkle. It may also cause alligatoring, checking, blistering, peeling, flaking, cracking, or scaling.

Another problem is applying a hard finish coat over a relatively soft undercoat, which can cause a paint to wrinkle and may result in alligatoring, checking, or peeling of the top coat.

Failing to brush out the paint can cause wrinkling, runs, or sags.

Improperly operating and incorrectly applying paint using spray equipment can cause several types of problems. Too much air pressure may result in trapping air in a paint or other organic coating, which can appear as bubbles visible beneath the surface. Permitting water in the air line used in spray applications can dilute the paint or other organic coating and cause it to cure improperly. Improperly adjusting a spraying device's pattern so that sprayed paint or other organic coatings are applied using too much air pressure can cause a wrinkled coating. Failing to remove oil and grease from lines can cause the problems mentioned earlier.

Failure to Protect the Installation

In addition to the measures mentioned under "Failure to Protect the Installation" in Chapter 7, there are other methods necessary to protect paint or other organic coatings before, during, and after installation. The types of errors that can lead to the failure of coatings include those that follow.

Damage may result from failing to protect applied paint or other organic coating materials from staining or marring during shipment or by contact with other construction materials before and during construction. One example is permitting contaminants, including airborne dust, such as from broom cleaning, to fall onto wet, newly painted or coated surfaces.

Allowing rain to fall on paint or another organic coating while it is still wet can pit the surface and remove the gloss from oil- or alkyd-based paints or clear organic coatings. It may even wash the paint or other coating partly—or even completely—off the metal surface.

Permitting the temperature in a space where paint or another air-drying organic coating has been applied to fall below 55 degrees Fahrenheit or rise above 95 degrees may create finish failures. High or low temperatures cause expansion and contraction in the paint or other air-drying finish and its substrates. Extremes may cause the paint or other organic coating to crack, especially if it is old or naturally hard or for some other reason not elastic enough to withstand the stresses involved.

Permitting deterioration of adjacent construction to occur to an extent that allows painted or coated metal to become wet or corrode is a further cause for possible failure.

Poor Maintenance Procedures

Failing to remove grease, dirt, stains, and other contaminants from paint and other organic coatings regularly can result in permanent stains.

The improper cleaning of paint and other organic coatings may damage them. Such coatings may be severely affected by sandpaper, steel wool, and abrasive or caustic cleaners, for example.

Failing to keep vines and other vegetation from growing on or close to building surfaces can also lead to damaged coatings on metal. Vines can grow into joints, carrying moisture with them and letting free water enter as well. This can lead to corrosion in concealed areas. Mildew and moss may grow on the wet surfaces and severely damage coatings.

Natural Aging

Over time, all paints and other organic coatings age and lose their properties. The manufacturer is the best source of data about the expected life of a given product. It should be kept in mind, however, that a manufacturer's projections may well assume the best of conditions. Paint and other organic coatings may fail, from any of the causes discussed in the following paragraphs.

Failing to repaint or recoat a surface at proper intervals may contribute to its complete eventual failure. Paints and other organic coatings, especially clear organic coatings that are exposed to sunshine, will fail faster than those in other locations. All will fail eventually if they are not maintained and repainted at proper intervals.

One form of failure is loss of elasticity. An old paint or other organic coating that has lost its elasticity becomes hard and is no longer able to respond to the expansion and contraction of the material to which it is applied. The result is crazing and, eventually, alligatoring, flaking, or peeling.

Natural erosion is another cause of failure. Weathering will slowly wear away paints and other organic coatings until eventually they can no longer properly protect a coated surface.

Evidence of Failure in Organic Coatings on Metal

In the following paragraphs, metal finishes' failures are divided into such failure types as "Rust or Other Corrosion." Each failure type is listed under its own heading. The list following each failure type indicates the location in this book where the causes of that failure type can be found.

A discussion of the errors that can lead to each failure cause listed can be found in the chapter whose number follows the listing, under a heading with the same name as the failure cause. To make them easier to locate, each of the failure causes listed is included in the Contents.

Thus, to discover the possible causes of a damaged finish on a metal fabrication, look under "Rust or Other Corrosion." One possible cause listed there is "Other Building Element Problems: Chapter 2." Open Chapter 2 to "Other Building Elements." Examine the problems discussed there that can lead to failure, to discover if any of them might be applicable to the type of damage being observed. If none of those reasons seem to correlate with the damage, look under the next possible cause and so on, until the actual cause has been identified. There may of course be more than one cause for a particular failure.

Rust or Other Corrosion

The appearance of rust or other corrosion on a metal surface may have been caused by many factors (Figs. 5-8 and 5-9). Refer to "Rust and Other Corrosion" in Chapter 4 for headings that address the causes of rust and other corrosion on surfaces without organic coatings. Most of the causes of rust and corrosion on metal surfaces that do have organic coatings will be found under one or more of the following headings:

- Other Building Element Problems: Chapter 2.
- Errors That Lead to Corrosion: Chapter 3.
- Errors That Lead to Fracture: Chapter 3.
- Bad Materials: Chapter 5.

Figure 5-8 The steel joint cover shown in this photograph rusted because it was not protected.

Figure 5-9 The rust and other stains on this wall occurred because the steel supports for the window air-conditioning unit were not protected from corrosion.

- Immediate Substrate Failure: Chapter 5.
- Improper Design: Chapter 5.
- Bad Workmanship: Chapter 5.
- Failure to Protect the Installation: Chapter 5.
- Poor Maintenance Procedures: Chapter 5.
- Natural Aging: Chapter 5.
- Improper Design: Chapter 6.
- Improper Fabrication: Chapter 6.
- Bad Installation Workmanship: Chapter 7.
- Failure to Protect the Installation: Chapter 7.

Stained or Discolored Paint or Other Organic Coatings

Blemishes include stains, brown spots and other discolorations, differences in color or sheen within a surface, mildew, and moss and other plant growth. The causes may be discussed under one or more of the following headings:

- Solid Substrate Problems: Chapter 2.
- Other Building Element Problems: Chapter 2.
- Errors That Lead to Corrosion: Chapter 3.
- Bad Materials: Chapter 5.
- Immediate Substrate Failure: Chapter 5.
- Improper Design: Chapter 5.
- Bad Workmanship: Chapter 5.
- Failure to Protect the Installation: Chapter 5.
- Poor Maintenance Procedures: Chapter 5.
- Natural Aging: Chapter 5.
- Improper Design: Chapter 6.
- Improper Fabrication: Chapter 6.
- Bad Installation Workmanship: Chapter 7.
- Failure to Protect the Installation: Chapter 7.

Wrinkling

Wrinkles appear in paint or other organic coatings when the surface dries before the underlying coating material has dried. This condition can result from one or more of the causes addressed under the following headings:

- Bad Materials: Chapter 5.
- Improper Design: Chapter 5.
- Bad Workmanship: Chapter 5.

Alligatoring and Checking

Cracks in paint or coatings that resemble an alligator's hide are called "alligatoring." Checking is similar, except that such cracks are smaller and not as noticeable. In their early stages these defects are also sometimes called crazing, crowfooting, or hairlining. They are caused by the inability of a top coat of a paint or other organic coating to adhere properly to underlying coats or the substrate. In their early stages these defects affect only the surface coat, but later the cracks may extend through to the substrate. They result from one or more of the following causes, discussed under their own headings:

- Concrete and Steel Structure Movement: Chapter 2.
- Metal Framing and Furring Problems: Chapter 2.
- Wood Framing and Furring Problems: Chapter 2.
- Solid Substrate Problems: Chapter 2.
- Other Building Element Problems: Chapter 2.
- Errors That Lead to Corrosion: Chapter 3.
- Bad Materials: Chapter 5.
- Immediate Substrate Failure: Chapter 5.
- Improper Design: Chapter 5.
- Bad Workmanship: Chapter 5.
- Failure to Protect the Installation: Chapter 5.
- Poor Maintenance Procedures: Chapter 5.
- Natural Aging: Chapter 5.
- Improper Design: Chapter 6.

Peeling, Flaking, Cracking, and Scaling

Peeling, flaking, cracking, and scaling may be advanced stages of alligatoring, checking, or blistering, or they may occur independently. Paint or other organic coatings may peel completely away from the substrate, or outer layers may peel off of undercoats (Fig. 5-10). When cracks extend completely through a coating's film down to the substrates, water works its way behind the film, causing it to flake (small pieces) or scale off. Scaling is an advanced form of flaking.

The problem shown in Figures 5-11 and 5-12 is a factory-applied-coating failure that resembles paint in its peeling, flaking, and scaling.

The failures addressed in this section may be due to one or more of the causes addressed under the following headings:

- Solid Substrate Problems: Chapter 2.
- Other Building Element Problems: Chapter 2.
- Errors That Lead to Corrosion: Chapter 3.
- Bad Materials: Chapter 5.
- Immediate Substrate Failure: Chapter 5.
- Improper Design: Chapter 5.
- Bad Workmanship: Chapter 5.
- Failure to Protect the Installation: Chapter 5.
- Poor Maintenance Procedures: Chapter 5.

Figure 5-10   The paint on this louver has peeled away from the shop coat, which usually indicates a preparation problem.

- Natural Aging: Chapter 5.
- Improper Design: Chapter 6.

Blistering

Blisters may appear as pimples, bubbles, pinholes, or pits in the paint or other organic coating. They may appear when either solvent or water vapor is trapped beneath a coating's impermeable surface. Blisters should not be confused with the white spots that appear in a lacquer coating over anodized aluminum, which are merely delaminations that do not usually contain vapor. There are several reasons that paint or other organic coatings may blister. Many of them are addressed under the following headings:

- Solid Substrate Problems: Chapter 2.
- Other Building Element Problems: Chapter 2.
- Errors That Lead to Corrosion: Chapter 3.
- Bad Materials: Chapter 5.
- Immediate Substrate Failure: Chapter 5.
- Improper Design: Chapter 5.

Figure 5-11 This photo shows a long-range view of a metal wall where part of the finish has failed. The finish has delaminated from the light areas near the top of the photo and collected in the dark areas near the bottom.

- Bad Workmanship: Chapter 5.
- Failure to Protect the Installation: Chapter 5.
- Poor Maintenance Procedures: Chapter 5.
- Natural Aging: Chapter 5.
- Improper Design: Chapter 6.

Excessive Chalking

When the resin in a cured paint or other organic coating disintegrates, chalking, which is a powder on the surface, is the result. Some chalking is desirable. It is, for example, the natural weathering method of chalking paints. When rain washes the chalking off, dirt and other soiling goes with it. Clear organic coatings that tend to weather by chalking rather than peeling are usually considered more desirable than those more likely to

Figure 5-12 This photo is a closer look at the upper portion of the wall shown in Figure 5-11. The missing areas of finish can be seen clearly.

peel. Excessive chalking can weaken an organic coating, however, as it becomes thinner, eventually resulting in a coating too thin to protect the underlying surface. Chalking in a material that is not supposed to chalk usually means that the material was defective. Excessive chalking may result from one or more of the following causes:

- Bad Materials: Chapter 5.
- Improper Design: Chapter 5.

Paint or Other Organic Coating Wash-off

When rain strikes an organic coating's surface before the material has cured (dried) properly, the outer layer, and sometimes the entire coating, may be washed away. This condition can result from one or more of the causes discussed in the following sections:

- Other Building Element Problems: Chapter 2.
- Bad Materials: Chapter 5.
- Bad Workmanship: Chapter 5.
- Failure to Protect the Installation: Chapter 5.
- Poor Maintenance Procedures: Chapter 5.

Paint Erosion

Paint wears away from the causes discussed under the following headings:

- Other Building Element Problems: Chapter 2.
- Failure to Protect the Installation: Chapter 5.
- Poor Maintenance Procedures: Chapter 5.
- Natural Aging: Chapter 5.

Failure to Dry as Rapidly as Expected

Paint and other field-applied organic coatings may remain tacky for extended periods and dry too slowly because of one or more of the causes discussed under the following:

- Other Building Element Problems: Chapter 2.
- Bad Materials: Chapter 5.
- Improper Design: Chapter 5.
- Bad Workmanship: Chapter 5.
- Failure to Protect the Installation: Chapter 5.

Failure to Ever Dry

Good paint or other organic coating materials will dry eventually, regardless of the application methods used or the prevailing climatic conditions. When a paint or other organic coating never dries it is for one of the reasons discussed under the following headings:

- Bad Materials: Chapter 5.
- Bad Workmanship: Chapter 5.

Poor Gloss

The failure of a paint or other organic finish to achieve or maintain its expected level of gloss may be due to one of the causes below:

- Other Building Element Problems: Chapter 2.
- Bad Materials: Chapter 5.
- Improper Design: Chapter 5.
- Bad Workmanship: Chapter 5.
- Failure to Protect the Installation: Chapter 5.
- Poor Maintenance Procedures: Chapter 5.
- Natural Aging: Chapter 5.

Telegraphing

Irregularities, foreign material, or soiling underneath a paint or other organic coating may telegraph through the coating or not be hidden properly due to one of the causes addressed under the following headings:

- Metal Framing and Furring Problems: Chapter 2.
- Wood Framing and Furring Problems: Chapter 2.
- Solid Substrate Problems: Chapter 2.
- Other Building Element Problems: Chapter 2.
- Bad Materials: Chapter 5.
- Immediate Substrate Failure: Chapter 5.
- Improper Design: Chapter 5.
- Bad Workmanship: Chapter 5.
- Failure to Protect the Installation: Chapter 5.
- Improper Design: Chapter 6.
- Improper Fabrication: Chapter 6.
- Bad Installation Workmanship: Chapter 7.

Runs and Sags

Paint or other organic coatings may not present a smooth surface, because of causes discussed under the following heading:

- Bad Workmanship: Chapter 5.

Cleaning and Repairing Paint or Other Organic Coatings on Metals

General Requirements

The extent of cleaning and repairing needed on paint or other organic coatings depends on the type and extent of the soiling or damage. This section discusses the cleaning and repairing needed when damage is only minor. "Applying Paint or Other Organic Coatings over Existing Surfaces" later in this chapter discusses major extensions of existing paint and other organic coatings, and the installation of new paint and other organic coatings over existing materials, including existing paint and other organic coatings. It also covers the removal of existing coatings and other preparation of substrates to receive new paint or other coatings.

The following discussion contains suggestions for cleaning and repairing paint and other organic coatings. Because the suggestions are meant to apply to many situations, they might not apply in a specific case. In addition, there are many possible situations that are not directly covered here. When a condition arises in the field that is not addressed here, advice should be sought from the additional sources of data mentioned in this book. Often, consultation with the manufacturer of the materials being cleaned or repaired will help. Sometimes it is necessary to obtain professional help (see Chapter 1). Under no circumstances should the specific recommendations in this chapter be followed without careful investigation and the application of professional expertise and judgment.

The discussion earlier in this chapter about new paint and other organic coatings applies as well to new paint or transparent finishes used to repair existing surfaces. There are, however, a few additional considerations to address when the new materials are applied in an existing space. The following discussion addresses these concerns. Some, but not all, of the requirements mentioned earlier are repeated here for the sake of clarity, but the reader should refer to earlier sections for additional data.

The requirements discussed later in this chapter under "Applying Paint or Other Organic Coatings over Existing Surfaces" apply also. Existing materials should be removed, substrates be prepared, and new materials applied in accordance with the paint or other organic coating material manufacturer's recommendations. Only experienced workers should be used to prepare for and apply paint or other organic coating materials.

Before an attempt is made to clean or repair existing paint or other organic coatings, the existing materials' manufacturers' brochures for products, installation details, and recommendations for cleaning and repairing should be available and referenced. It is necessary to be sure that the manufacturers' recommended precautions against using materials and methods that may be detrimental to the materials are followed. It is impossible to emphasize too strongly the need to have access to such data when existing coatings are of the baking-finish type or are transparent coatings over decorative finishes such as anodized aluminum or copper alloys.

It is difficult if not impossible to actually repair paint or other organic coatings in the true sense of the word, especially baking-finish coatings and transparent protective coatings on aluminum and copper alloys. Small patches and repairs are almost always obvious and for that reason are seldom acceptable. Therefore, in this chapter the word "repair" with regard to paint or another organic coating almost always means to remove or properly prepare an existing material and cover it with a new coat of material.

Areas where repairs are to be made should be inspected carefully to verify that existing materials that should be removed have been and that the substrates and structure are as expected and not damaged. Sometimes substrate or structure materials, systems, or conditions are encountered that differ considerably from those expected. Unexpected damage may be discovered. Both damage that was previously known and damage found later should be repaired before paint or other organic coatings are repaired. Cleaning should be done, substrates prepared, and repairs made strictly in accordance with the materials manufacturers' recommendations. Repairs should be made only by experienced workers.

Materials to be used in making repairs should not be permitted to freeze. Work spaces should be kept above 55 degrees Fahrenheit before, during, and after repairs are made.

Adequate ventilation should be provided while the work associated with the preparation and application of paint or other organic coating materials is in progress. This may be somewhat more difficult when operating in an existing building, especially one that is occupied. The paint or other organic coating materials manufacturers' recommended safety precautions should be followed.

Cleaning Soiled Surfaces That Do Not Need Repair

Refer to "Cleaning and Maintaining Existing Metal Surfaces" in Chapter 4 for a discussion of cleaning existing metal surfaces having mechanical, chemical, and other inorganic finishes. Also see additional discussion under "Cleaning of Existing Metal Fabrications" in Chapter 7.

Existing surfaces to be cleaned but not repainted or recoated should have soiling, stains, dirt, grease, oil, and other unsightly contaminants completely removed.

Dirt, soot, pollution, cobwebs, insect cocoons, and the like can often be removed from exterior surfaces using a water spray, followed by scrubbing with a mild solution of household detergent and water. Cleaned surfaces should be rinsed thoroughly with clean water, then allowed to dry.

Cleaning should be done in strict accordance with the instructions of the paint or other organic coating manufacturer. Detergents and cleaning compounds should be compatible with existing finishes on both the metal and adjacent materials. Abrasives, acids, or caustic cleaning agents or devices should not be used. The materials and methods used should be those recommended by the paint or other organic coatings' manufacturer.

Surfaces damaged from using incorrect products or from improper use of cleaning agents should be refinished.

Structural Framing, Substrates, and Other Building Elements

Though improperly designed or installed or damaged concrete or steel structures, steel or wood framing and furring, solid substrates, or other building elements, which are discussed in Chapter 2, may be responsible for paint or other organic coatings' failure, repairing them is beyond the scope of this book. They should be investigated as a possible source of the failure and repaired as necessary. This chapter assumes that when such items have been the cause of failure they have been repaired and will now serve as satisfactory support for the materials being repaired.

Materials and Manufacturers

Paint and other organic coatings and associated materials used in making repairs should be as listed in the part of this chapter called "Organic Coating Materials." The discussion there about manufacturers also applies here.

Not only must new materials be compatible with each other within each paint or other organic coating system, as is necessary in new buildings, but new materials for use in existing buildings must also be selected for their compatibility with existing materials that will remain in place. Tests should be conducted in the building to ensure compatibility, and incompatible materials should not be used.

Materials should be delivered to a project site in the manufacturer's original packaging. Materials should be packed, stored, and handled carefully to prevent damaging them.

Paint and Other Organic Coating Systems

The discussion of paint and other organic coating systems earlier in this chapter applies generally to systems to be used in making repairs, though there are differences. The new materials should match those in the original application, but some coats may be eliminated. For example, a primer coat may not be necessary when the damage affects only finish coats and can be eliminated without removing or damaging existing undercoats.

Surfaces Usually Repaired

Any paint or other organic coating may need repair. Where major damage is present, the discussion under "Applying Paint or Other Organic Coatings over Existing Surfaces" applies. Here we will discuss only minor repairs to previously coated surfaces. Making even minor repairs to paint or other organic coatings usually means that part of the original coating must be removed and new materials applied.

Preparation for Repairing Paint or Other Organic Coatings

Surfaces to be repaired should be prepared as discussed later under "Applying Paint or Other Organic Coatings over Existing Surfaces." The requirements suggested there are generally applicable here also.

Before repairs are attempted, the proposed technique and materials to be used should be tested in an inconspicuous area for compatibility of materials and matching of the existing color and finish.

It is necessary to conduct an inspection of the surface before a decision can be made about whether to repair a damaged surface or repaint or recoat it. Some testing may also be necessary. Although some painting contractors are able to conduct such examinations and tests, many are not, especially when the coating is a baking finish or a clear coating over aluminum or a copper alloy. Most national paint manufacturers employ personnel trained and equipped to make inspections and tests and recommend methods of dealing with paint failure. Some can also deal with failures in baking-finish organic coatings. It will probably be necessary, however, to contact the metal's producer for assistance with a clear coating over aluminum or a copper alloy.

Tests that may be required include ones for compatibility, coating thickness, and moisture content. The results of these tests will often point directly to the cause of the failure. Test results can also help prevent further failure caused by misdiagnosis or by using the wrong methods to make repairs or to repaint or recoat failed surfaces. Examination may show, for example, whether just a surface coat is peeling or whether the entire thickness of the paint or other organic coating system is affected. Microscopic examination may be necessary to determine the number of coats already applied. Chemical testing may also be needed, to determine the type of organic coating—oil-based or water-based paint, for example—used in existing coats.

Even before tests are made, it may be possible to determine the cause of some paint and other organic coating failures. Where paint peeling or chipping has occurred, for example, a piece of paint can be examined. Slick surfaces may mean that the substrate was too smooth and must be sanded or otherwise roughened before new paint will adhere properly. The back of a particular paint chip could also be slick if the substrate was coated with oil or grease when the paint was applied. A chalky residue on the back of paint that was on metal may be rust. White rust will form even on zinc-coated metal. And paint that was over a galvanized surface coated with a waxy film may indicate that an oil-based primer was used.

When blistering has occurred, a diagnosis can be made by slitting a blister. If the bare substrate is visible, the blister was probably caused by trapped moisture vapor. If instead there is coating material beneath the blister, the problem was caused by solvent vapor that could not escape into the atmosphere because the coating's surface skin formed too rapidly.

Repairing, Cleaning, Clean-up, and Protection

The requirements discussed later in this chapter under "Applying Paint or Other Organic Coatings over Existing Surfaces" generally apply also to repairs.

Existing paint or other organic coatings that are to be repainted or recoated should first be cleaned properly. For example, paint should be scrubbed using household cleaner and water, rinsed thoroughly with clean water, and allowed to dry before repainting it. Mildew, stains, and other blemishes should be treated as discussed under "Applying Paint or Other Organic Coatings over Existing Surfaces."

The extent of the new paint or other opaque organic coating that is applied during a repair may be just the repair itself, but more normally it will extend to natural breaks, such as corners or projections. Sometimes a metal item with a baking finish cannot be successully recoated in the field, and must be removed and sent back to the shop for refinishing. Touching up small areas in such surfaces is often possible, however, by using air-drying repair coatings that successfully match the original coating. Until relatively recently, fluorocarbon polymers were difficult to repair successfully in the field, but now there is at least one air-drying fluorocarbon polymer resin-based repair material available.

Repaired surfaces with clear coatings are usually refinished completely. Where a surface cannot be refinished without affecting its edges and projections, these areas are usually also refinished. Some clear finishes are difficult to remove without damaging their underlying mechanical or chemical finishes. They must often either be just patched, which may be unsightly, or the entire item must be removed and shipped to a shop for refinishing.

What follows are recommendations for dealing with common paint and other organic coating failure problems. They should, of course, be adapted to suit actual conditions. Although these suggestions for repairing paint are by and large universally true, the procedures recommended for other organic coatings, especially baking-finish coatings and clear coatings, are so specific to the individual products that it is impossible to make statements that apply universally to them all. In every case, the advice of the coating manufacturer and the metal item producer should be sought before attempting to repair either of those types of coatings. The same is true for special types of organic coatings, such as fluorocarbon polymers.

Brown Spots on Painted Surfaces. The recommended method of repair for brown spots depends on the cause of the spots. When they are caused by using a latex paint where high humidity is present, for example, the spots can usually be cleaned away with a damp sponge. Abrasive cleaners should not be necessary, and repainting will usually not be needed.

When the brown spots are inherent in the paint or coating, they may resist removal and necessitate complete repainting or recoating of the surface. Tests should be made first, however, to ensure that the spots are not a chemical problem that will also extend through the new finish. In such a case it may be necessary to remove the existing coating down to bare metal and start all over again.

Mildew. It is first necessary to establish that an observed stain is actually mildew. A simple test is to place a drop of household bleach on the stain. Mildew will turn white, dirt will not. Mildew can be removed by using the formulation mentioned earlier in this chapter under ''Preparation for Paint and Other Organic Coatings'' in the section ''Preparing a Site for Field Application of Paint or Other Organic Coatings.'' Preventing a recurrence may be more difficult, however. Mildew-resistant paints may help. The most effective prevention technique is to remove whatever caused the shade and dampness that permitted the growth in the first place. For example, vegetation may be trimmed to permit sunlight to strike the surface. Sources of water should be controlled or removed. Adding a rain gutter and downspout system will sometimes help.

Excess Chalking. Slight chalking can be removed by lightly brushing the surface. Sometimes light chalking on paint will succumb to washing with a household detergent and water and soft brushes. Where such is permitted by law, heavy chalking can be removed by scrubbing with a solution of trisodium phosphate and water, using a stiff brush. Care should be taken not to damage the surface.

Wrinkles in Paint or Other Coatings. When a condition is minor, it may be possible to scrape and sand down the wrinkled coat and repaint it. More severe conditions may require the complete removal of all paint down to the bare substrate before repainting.

Wrinkles in baking-finish coatings and clear coatings will probably require completely refinishing the surface. If an air-drying coating is available that matches a wrinkled baking-finish coating, it may be possible to treat that surface as described above for wrinkled paint.

Air-drying materials are available for repairing or recoating most clear finishes, but such repairing may be apparent. Some clear coating materials dissolve themselves, which makes small repairs easy to make. Large repairs, however, may be difficult and may force complete removal of an existing clear coating before a new one can be applied successfully.

Crazing.   An early stage of alligatoring, crazing in paint can be sanded and repainted without complete removal of the paint. Its tiny cracks will be apparent on examination but will be filled with new paint and will, therefore, protect the substrate. If the final appearance is not acceptable, complete removal of the crazed paint will be necessary.

The same procedure applies to a baking-finish coating if an air-drying patching material is available to match the baking finish. If not, the item must be refinished, which will probably require returning it to the shop.

Crazed clear finishes must normally be completely refinished, but some can be repaired.

Alligatoring and Checking.   Because checking and alligatoring often affect only the surface coats, some sources say that it is sometimes possible to sand or scrape away only the affected coats before recoating. Although this process may be satisfactory for paints on metals, it may not work on a baking finish, and it almost certainly will not work on a clear finish. Even in paint, when cracks extend completely through the paint, complete removal is always necessary.

Cracking and Scaling of Paint.   Cracking and scaling affect the entire thickness of a coating, which makes it necessary to completely remove a coating that has this kind of damage and recoat the exposed bare metal. This type of defect in paint is often caused by the paint's being too brittle for the surface; thus, the new paint should be more flexible in nature.

Blistering.   Blistered paint or other organic coatings in which the blisters were caused by moisture on the substrates must be removed completely, down to the bare substrates. The substrates must then be allowed to dry before they are repainted or recoated.

Blistered paint caused by solvent vapors may be removed only down to sound paint.

Peeling.   When a paint or other organic coating has peeled away from the substrates, complete removal and repainting or recoating is necessary. If the peeling coating is a baking finish, it will be necessary to return the affected item to the shop for refinishing, unless an air-drying coating is acceptable.

When a surface coat has peeled away from its undercoats, it may be possible to remove only the peeling coat down to sound paint. When the peeling was caused by the use of incompatible paints, an incompatible top coat or coats should be completely removed.

Runs and Sags. For runs and sags, proper application is the best cure. Runs and sags in paint that have been permitted to dry can be removed with sandpaper. Another coat of paint may then be necessary to achieve the desired finish.

Runs and sags in baking-finish coatings and clear coatings may not be repairable without refinishing.

Excessive Paint Buildup. Existing paint with a total film thickness of more than 1/6 inch should be completely removed.

Naturally Aged Paint and Other Organic Coatings. When paint and other organic coatings have reached their expected life spans, repainting or recoating is necessary. When they are permitted to begin to fail, re-painting or recoating then becomes much more difficult. In some cases, particularly with clear coatings, complete removal of an existing finish may be necessary. With baking finishes the item must be returned to the shop for recoating, if the original finish is to be truly duplicated. This is seldom done, however; such items are usually painted in place with an air-drying paint or other organic coating.

Clear Coatings over Other Metal Finishes That Are Damaged. Clear organic coatings are often applied over anodized aluminum, and sometimes over copper alloys, to protect their finishes. When the underlying metal or its mechanical or chemical finish has become damaged, it is often necessary to remove the clear organic coating to repair the underlying material or finish.

Nonanodized or clear anodized aluminum may begin to oxidize and turn dark. Before the aluminum can be cleaned, as described in Chapter 4, any organic coating covering it must be removed. If reanodizing is re-quired, the item must be shipped to an anodizing plant. After the aluminum has been cleaned, and reanodized if desired, an organic coating (usually methacrylate lacquer) may be applied to protect the new finish. For the reasons mentioned earlier, however, color-anodized aluminum should not have a methacrylate lacquer applied, regardless of the extent of the cleaning or repairs. Another possible solution is to skip the lacquer and give the newly finished aluminum a coat of wax or oil (see Chapter 4). The wax or oil will need to be renewed more frequently than would lacquer.

Copper alloys with a clear coating, such as lacquer, may begin to appear dull over time. If so, their luster can often be restored by removing the clear coating and refinishing the copper alloy. For example, if the copper alloy is a lacquered brass, the lacquer can be removed with amyl nitrate (banana oil), denatured alcohol, or acetone. Other clear coating types will require different removal methods. The finish may also need restoration. The method employed will depend on the type of finish. No attempt should be made to restore the finish on a copper alloy unless the exact finish is known. Only then should the restoration be made, in accordance with the metal manufacturer's and finisher's recommendations. Restoration may be simple, as in the case of tarnished brass, which can be restored with a common brass cleaner, or complicated, as for a faded patina on bronze. After the finish has been restored, a new clear organic coating, either of the original or a different type, will probably be recommended by the metal finisher. Sometimes, however, a wax or oil coating will be recommended (see Chapter 4).

Applying Paint or Other Organic Coatings over Existing Surfaces

The following requirements contain some suggestions for applying new paint and other organic coatings over existing materials. Discussion of baking-finish coatings has been omitted here, because items with such coatings that are so severely damaged as to require recoating are either refinished in place with paint or are removed and returned to the shop for refinishing. In the first case, the discussion here applies. In the second case, the methods used are so specialized and depend on so many variables that they are not usually subject to control by building owners, architects, engineers, or general building contractors.

Because the suggestions that follow are meant to apply in many situations, they may not apply to a specific one. In addition, there are many possible cases that are not specifically covered here. When a condition arises in the field that is not addressed here, advice should be sought from the additional sources of data mentioned in this book. Often, consultation with the manufacturer of the materials being applied will help, but sometimes it may be necessary to obtain professional help (see Chapter 1). Under no circumstances should these recommendations be followed without careful investigation and the application of professional expertise and judgment.

The discussion earlier in this chapter about new paint and other organic coating materials and their installations on new surfaces also applies generally to paint and other organic coatings installed on existing surfaces. There are, however, a few additional considerations to address when such materials are applied in an existing space. Some but not all of the requirements discussed earlier are repeated here for clarity, but readers should refer to the earlier discussion for more detailed data.

Substrates should be prepared and materials applied in accordance with the paint or other organic coating material's manufacturer's recommendations. Only experienced workers should be used to prepare for and apply paint or other organic coating materials.

Materials manufacturers' brochures including installation, cleaning, and maintenance instructions should be collected for each type of paint and other organic coating material and kept available at the job site.

Adequate ventilation should be provided while work associated with paint or other organic coating materials preparation and application is in progress. This may be somewhat more difficult in an existing building, especially one that is occupied. The paint or other organic coating materials' manufacturers' recommended safety precautions should be followed.

Areas where repairs are to be made should be inspected carefully to verify that existing materials that should be removed have been removed and that the substrates and structure are as expected and not damaged. Sometimes substrate or structure materials, systems, or conditions are encountered that differ considerably from those expected. Sometimes unexpected damage is discovered. Both damage that was previously known and damage found later should be repaired before new paint or other organic coating materials are applied.

Materials and Manufacturers

Paint and other organic coating and associated materials should be as discussed earlier in this chapter under "Organic Coating Materials," and the discussion there about manufacturers and products applies here.

In addition to having materials be compatible with each other within each paint or other organic coating system, as is necessary in new buildings, the materials for use in existing buildings must also be selected for their compatibility with existing materials that are to remain in place. Tests should be conducted in the building to ensure compatibility, and incompatible materials should not be used. For example, applying water-based paints over solvent-based ones can create serious problems. The water-based paint may lift the oil-based one from its substrate or refuse to bond with it. All paint applied before 1950 is likely to have an oil base, since water-based paints were unheard of before then. Even today, except in residences, most paint applied on metal is oil or alkyd based.

Some materials are limited in how effective they are for repainting or recoating existing surfaces. Some transparent finishes containing urethane resins, for example, are not applicable over other finishes, and maybe not even over a previous coating of the same material, without extensive preparation. Lacquer cannot be used over solvent-based paints, because the lacquer will act as a paint remover.

Materials should be delivered to the project site in the manufacturer's original packaging. Materials should be packed, stored, and handled carefully to prevent damage.

Paint and Other Organic Coating Systems

The discussion of paint and other organic coating systems earlier in this chapter in "Paint and Other Organic Coating Systems" under "Organic Coating Materials" applies to systems to be used in existing buildings, but with several differences. The materials themselves are basically the same, but some coats may be eliminated when the existing paint or other organic coating is sound. It is also necessary to ensure that the new paint and finish systems selected are appropriate for repainting or refinishing every existing surface under every condition to be encountered.

Preparation for Applying Paint or Other Organic Coatings

Areas to be repainted or recoated, and the conditions under which paint or other organic coatings are to be applied, should be inspected carefully. It is not unusual to find conditions other than those expected. Unsatisfactory conditions should be corrected before work related to paint or other organic coatings is begun.

Some testing of existing surfaces may be necessary. Some painting contractors are able to conduct such examinations and tests, but many are not. Most national paint manufacturers employ personnel trained and equipped to make inspections and tests and recommend methods of dealing with paint failures. If examining or testing clear coatings is deemed necessary, the manufacturer of the coating or the metal manufacturer or finisher will probably have to be asked to do it.

Tests that may be required include ones for compatibility and determining existing coating thicknesses and moisture content. The results will often point directly to the cause of a failure. Tests' results can also help prevent further failures from misdiagnosis or using the wrong methods to make repairs or recoat failed surfaces. Examination may show, for example, whether just a surface coat of paint is peeling or the entire thickness is affected. Microscopic examination may be necessary to determine the number of coats already applied. Chemical testing may also be needed, to determine the type of paint (oil or water based, for example) used in existing coats. Tests may be necessary to determine the type of a clear coating.

Paint and Other Organic Coating Removal. Where existing paint is essentially sound, it is necessary to remove only loose and peeling paint, rust, oil, grease, dust, soiling, and other substances that would affect the bond or appearance of new paint. Exceptions must be made, however, when there is a compelling reason to remove more paint that is not associated with the paint itself and when the total thickness of the existing paint is 1/16 inch or more. It is not necessary to remove paint from surfaces that have only minor blemishes, dirt, soot, pollution, cobwebs, insect cocoons, and the like unless removal is needed to make repairs to substrates. Such foreign matter may be removed using a water spray, followed by scrubbing with a mild solution of household detergent and water. The cleaned surfaces should be rinsed thoroughly with clean water and then permitted to dry.

Where the existing paint displays crazing, blistering, peeling, or cracking of the top layer or top few layers only, it may be possible to remove existing paint only down to sound layers. It will, however, be necessary to remove existing paint down to bare substrate and correct an underlying problem when existing paint displays excessive chalking, or blistering, peeling, flaking, cracking, or scaling through its entire thickness. It may also be necessary to remove existing paint when the introduction of new substrate material prohibits a smooth transition from existing to new paint. Existing paint that has been allowed to build up to more than 1/16 inch in thickness should also be removed down to the bare substrate.

The coating material's manufacturer's and metal finisher's advice should usually be followed when deciding whether an existing clear coating must be completely removed or not. If complete removal of paint or another organic coating is necessary, it should be accomplished by one of the following methods.

Limited Paint Removal. Where paint removal is limited to small areas, abrasive methods should be used, including scraping and sanding. Hand tools should be used in most cases. Gouging the substrates should be avoided. Mechanical abrasive methods such as orbital sanders and belt sanders may be appropriate in some circumstances but should be used carefully to prevent removing too much material. Rotary sanders, sandblasting, and waterblasting are not appropriate for limited paint removal and should not be used.

Total Paint Removal Using Heat. For total paint removal, thermal methods such as an electric heat plate or electric heat gun should be used.

However, blowtorches and other flame-producing methods are dangerous and should not be used for removing paint. Fire-insurance requirements should be checked and proposed methods cleared with the insurance provider before any hot-process removal is undertaken.

Total Paint Removal Using Chemicals. Chemicals formulated especially for the purpose, such as solvent-based strippers and caustic strippers, may also be used for total paint removal. Such materials must be used carefully following the manufacturer's directions exactly. Appropriate precautions are necessary to protect workers and others from inhaling vapors, fire, eye damage, chemical poisoning from skin contact with chemicals, and other dangers associated with using chemicals. Lead residues and other harmful substances should be disposed of properly. Chemical strippers must be removed completely before paint is applied. The residue left by some strippers will impede the adhesion of new paint.

Total Paint Removal Using Abrasives. Abrasive methods include scraping, sanding, waterblasting, sandblasting, and blasting with other abrasives. One abrasive product that has been used successfully in some conditions is made from corncobs. Hand tools should be used where appropriate. The gouging of substrates should be avoided. Such mechanical abrasive methods as orbital sanders and belt sanders may be appropriate in some circumstances, but such devices should be used carefully to prevent removing too much material. Sandblasting is probably not an appropriate means of removing paint from galvanized metal, because preventing damage to the metal while doing so requires more expertise than is commonly available and sandblasting removes the galvanizing.

Clear Coatings. Removal of clear coatings should be done in strict accordance with the instructions of the coating manufacturer and metal producer.

Hazardous Materials. The improper removal and disposal of existing chromium- and lead-based paints can create serious health and legal problems. Removal and disposal of such materials is regulated at the national level and by many state and local jurisdictions as well. For example, material containing more than 5 parts per million of lead or chromium is classified by the Environmental Protection Agency (EPA) as hazardous waste and is subject to all applicable regulations. Serious problems can also occur when materials to be removed contain asbestos, arsenic, or other toxic substances. Even methylene chloride, which is used in many commercial paint removers, is classified as a toxic waste material. Removal and disposal of such materials must be done in strict accordance with laws and recognized safety precautions. Regardless of who is at fault, stiff fines and penalties may be imposed on owners, architects, and contractors if such materials are handled or disposed of improperly. A complete discussion of hazardous materials removal and disposal is beyond the scope of this book. When faced with such a problem, obtain copies of the applicable rules and regulations and follow them explicitly. The Steel Structures Painting Council (see the Appendix) is a source for current recommendations about cleaning hazardous materials from metals. Their recommendations may also be applicable to other materials or will at least suggest other sources. The actual removal of hazardous wastes is best done by professionals expert in such work.

Existing Surface Preparation. The preparation and cleaning procedures for each specific substrate condition should follow the instructions of the paint or other organic coating material's manufacturer.

When existing finishes have been removed, their removal should be complete, including undercoats and other materials that would affect the new paint or coating or show through new materials. The cleaned substrates should be completely free from films, coatings, dust, dirt, and other contaminants.

Where paint or other existing organic coatings have been completely removed, the substrate should be prepared as it would be if new.

Hardware, accessories, machined surfaces, plates, fixtures, and similar items not to be painted or coated should be removed before the surface preparation for painting or another organic coating begins. Alternatively, such items may be protected by surface-applied tape or some other type of protection. Even when such items are to be painted or coated, they are often removed to make painting or coating them and adjacent surfaces easier. After the painting or other coating has been completed in each area, the removed items can be reinstalled.

Surfaces should be cleaned before the first paint or other organic coating is applied, and if necessary between coats to remove oil, grease, dirt, dust, and other contaminants. Activities should be scheduled so that contaminants will not fall onto wet newly painted or coated surfaces. Paint or other organic coatings should not be applied over dirt, rust, scale, oil, grease, moisture, scuffed surfaces, of under other conditions detrimental to the formation of a durable paint or other coating film.

Mildew should be removed from painted surfaces and neutralized. Some sources recommend scrubbing affected surfaces thoroughly with a solution made by adding two ounces of a trisodium phosphate-type cleaner and eight ounces of sodium hypochlorite (Clorox) to one gallon of warm water. This chemical may not be legal in every jurisdiction, however. If necessary, a scouring powder may be used to remove mildew spores, but care must be taken to prevent damage to the surface being cleaned. Cleaned surfaces should be rinsed with clear water and allowed to dry thoroughly. Existing surfaces containing natural salts that have not been washed away because rain does not strike the surface should be washed. The usual method is to wash such surfaces with a solution of trisodium phosphate to remove the salts. Where that material is not legal, another must be used. Susceptible areas are exterior surfaces under eaves and beneath roof overhangs.

Existing surfaces to be repainted or recoated that are glossy and do not require complete removal of an existing coating should be roughened by sanding or another appropriate method that will produce enough surface tooth to accept the new paint or other coating.

Excess chalking can be removed from existing sound paint with a solution of one-half cup of household detergent to a gallon of water applied by a medium-soft bristle brush. After the chalking has been removed, the surface should be rinsed with a clean-water spray and permitted to dry thoroughly. New paint should be applied before chalking starts again.

When paint or another organic coating is stained, the source of the stain should be located and the conditions that produced it be corrected. When the staining is caused by rust or other corrosion, the metal should be uncovered, hand sanded, and coated with a rust-inhibitive primer or other coating to prevent further corrosion.

Shoulders at the edges of sound existing paint or other organic coatings should be ground smooth and sanded as necessary to remove the shoulders so that flaws will not telegraph through the new paint or other organic coating. Where recommended by the paint or other opaque organic coating manufacturer, such edges may be feathered with drywall joint compound or another method recommended by the paint or coating manufacturer.

Existing paint to be repainted should be scrubbed using household cleaner and water, rinsed thoroughly with clean water, and allowed to dry before painting.

Before the application of paint or another organic coating is started in an area, that space should be swept clean with brooms and excessive dust removed. After a painting or coating operation has been started, however, that area should not be swept with brooms. Necessary cleaning should then be done using commercial vacuum-cleaning equipment.

Surfaces to be painted or coated should be kept clean, dry, smooth, and free from dust and foreign matter that would adversely affect adhesion or appearance.

Even when surfaces cannot be put in proper condition by customary cleaning, sanding, and puttying, proper conditions for applying paint or coatings must still be created, using extraordinary methods if necessary.

Existing ferrous-metal surfaces that are not galvanized, shop coated, or previously painted and are to be painted should be cleaned of oil, grease, dirt, loose mill scale and other foreign substances, using solvent or mechanical cleaning methods as recommended by the paint manufacturer. The recommendaions of the Steel Structures Painting Council should be followed, unless they conflict with the paint manufacturer's recommendations. Bare and sandblasted or pickled ferrous metal should be given a metal treatment wash before a primer is applied.

It is necessary to remove dirt, oil, grease, and defective paint from shop coated or previously painted ferrous, galvanized, and nonferrous metal surfaces using a combination of scraping, sanding, wirebrushing, sandblasting, and by applying a cleaning agent recommended by the paint manufacturer. Surfaces should be wiped clean before they are painted. Shoulders should be sanded if necessary to prevent telegraphing. Where necessary to produce a smooth finish, existing paint should be stripped to the bare metal.

Removing rust and other corrosion before painting or repainting is essential. The methods used will depend on the amount of rust or other corrosion present and the severity of the problem. The normal methods include sandblasting, power tool cleaning, hand cleaning, and phosphoric-acid wash cleaning. In each case the recommendations of the Steel Structures Painting Council should be followed for cleaning ferrous metals and those of the manufacturer for nonferrous metals.

Successfully painting existing unpainted galvanized metal is one of the more difficult painting problems. For one thing, authoritative sources disagree about the proper methods and materials to be used to prepare galvanized metals for receiving paint. Most sources do agree, however, that existing galvanized surfaces will contain white rust. And if the galvanizing is damaged or severely corroded, the galvanized metal may also contain common rust. Both types of rust and any other contaminants present must be removed before paint is applied. The methods for removing rust include wirebrushing, sanding, and blasting. The brush-off blast-cleaning method recommended by the Steel Structures Painting Council is often used.

Existing galvanized surfaces may collect oils and grease, dirt and dust, smoke particles, soot, and other pollutants and contaminants. They may also still have on their surfaces the residue of oils, waxes, silicones, or silicates applied during fabrication. Such materials must be completely removed before paint is applied. Some sources say that galvanized surfaces should be cleaned using mineral spirits or xylol. Others say to use a solvent wash and specifically recommend against using mineral spirits. Some recommend using petroleum spirits, but others argue against them.

After all the oils, white rust, and other contaminants have been removed, bare galvanized surfaces should be pretreated, according to some sources. Some say to use a diluted vinegar solution, which is a weak acetic acid. Others recommend a proprietary acid-bound resinous or crystalline zinc-phosphate preparation or phosphoric acid. Still others state that vinegar and other acids should not be used on galvanized metals, because they are said to remove some of the galvanizing and leave a residue that can cause an applied paint to peel.

Before a new primer paint is applied, damage to galvanizing should be repaired, using galvanized repair paint manufactured for the purpose.

The best way to determine the proper preparation methods to use on existing unpainted galvanized metals is to ask the paint's manufacturer for recommendations, and then follow them. Then there will at least be someone to complain to if paint failure occurs.

Bare aluminum should be washed with mineral spirits, turpentine, or lacquer thinner, as appropriate. It should then be steam cleaned and wire-brushed to remove loose coatings, oil, dirt, grease, and other substances that would reduce bond or harm new paint, and to produce a surface suitable to receive new paint.

Bare stainless steel should be washed with a solvent and prepared according to the paint manufacturer's recommendations.

Bare copper should have stains and mill scale removed by sanding. Then dirt, grease, and oil should be removed with mineral spirits. If a white alkyd primer was used, the copper should be wiped clean, using a clean rag wetted with Varsol.

Bare lead-coated copper should be sanded down to bright metal. Grease and oil should be removed, using a solvent recommended by the metal's manufacturer.

Bare terne-coated metal should be cleaned with mineral spirits and wiped dry with clean cloths. Immediately after cleaning, surfaces should be given a coat of a pretreatment specifically formulated for use on terne-coated metal. It should contain 95 percent linseed oil. The pretreatment should be allowed to dry at least 72 hours or longer if necessary to ensure that it is thoroughly dry before primer is applied.

Just as welds, abrasions, factory- or shop-applied prime coats, and shop-painted surfaces in new materials should be touched up after erection and before calking and application of the first field coat of paint, previously painted surfaces should also be touched up. On new surfaces the touch-up material should be the same as the primer. On previously painted surfaces the touch-up paint should be the same material as the first field coat. Surfaces that are to be covered should be touched up before they are concealed. Exposed touch-ups should be sanded smooth.

Paint and Other Organic Coating Application

The type of material for each coat, location, and use should be one designated for the purpose by the paint or other organic coating material's manufacturer, as stated in their published specifications.

Water-based paints should be applied only when the temperature of the surfaces to be painted and of the surrounding air is between 50 and 90 degrees Fahrenheit, unless the paint manufacturer's printed instructions say otherwise.

Solvent-thinned paints and other organic coatings should be applied only when the temperature of the surfaces to be coated and of the surrounding air is between 45 and 95 degrees Fahrenheit, unless the coating manufacturer's printed instructions recommend otherwise.

Paint and other organic coatings should not be applied in snow, rain, fog, or mist; when the relative humidity exceeds 85 percent; to damp or wet surfaces; or to extremely hot or cold metals unless the material manufacturer's printed instructions recommend otherwise. Applying paint or other organic coatings to surfaces exposed to hot sun should be avoided. Paint and other air-drying organic coatings may be applied during inclement weather, however, if the areas and surfaces to be painted or coated are enclosed and heated to within the temperature limits specified by the paint or other coating's manufacturer during the application and drying periods.

Once the application of paint or other organic coatings has been started within a building, a temperature of 65 degrees Fahrenheit or higher should be provided where the work is being done. Wide variations in temperatures, which might result in condensation on freshly coated surfaces should be avoided.

Paint and other organic coating materials should be mixed, thinned, and applied in accordance with the manufacturer's latest published directions. Materials should not be thinned unless the manufacturer specifically recommends doing so. The applicators and techniques should be those best suited for the substrate and the type of material being applied.

Sealers and undercoats should not be varied from those recommended by the paint or other organic coating's manufacturer.

Materials not in actual use should be stored in tightly covered containers. Containers used for storage, mixing, and application of paint or other organic coating materials should be maintained in a clean condition, free of foreign materials and residues.

Paint or coating materials should be stirred before application, to produce a mixture of uniform density, then stirred again as required during application. Surface films should not be stirred into the material. Film should be removed and, if necessary, the material strained before it is used.

A new paint or other organic coating applied on an existing, previously painted or coated surface may extend to the entire surface area or only part of it, depending on the requirements of the project and the recommendations of the manufacturer. When only partial repainting or recoating is required, it is usually best to extend repainting or recoating to natural breaks, such as corners or projections.

Repainting or recoating of existing surfaces should not be started until patchwork, extensions, repair work, and new work in the space has been completed.

When a flat paint finish is applied over a previously painted or coated surface, it is usually best to apply all the coats recommended by the paint manufacturer for that paint system for new work.

When a semigloss or gloss paint finish is to be applied over a previously painted or coated surface, the first coat (primer) should be applied in accordance with the paint manufacturer's specifications and the final coats should be applied as they would be on a new surface.

Before applying new paint or other organic coatings over existing paint, other organic coatings, or another existing finish, it is best to apply the paint or other organic coating in small inconspicuous locations representative of each condition that will be encountered, to test for compatibility.

Workmanship should be of a high standard. Paint and other organic coatings should be applied by skilled mechanics using the proper types and sizes of brushes, roller covers, and spray equipment. The equipment used should be kept clean and in proper condition. Most paint that is field applied to metal is put on with brushes, though rollers and spray equipment are sometimes used. The same rollers or brushes used to paint concrete or masonry should not be used on metal. Rollers used for a gloss finish and the corresponding primer should have a short nap.

Paint and other organic coating materials should be applied evenly and uniformly, under adequate illumination. The surfaces should be completely covered and should be smooth and free from runs, sags, holidays, clogging, and excessive flooding. Completed surfaces should be free of brush marks, bubbles, dust, excessive roller stipple, and other imperfections. Where spraying is either required or permitted, paint should be free of streaking, lapping, and pileup.

The number of coats recommended by the manufacturer for each particular combination of use, substrate, and paint or other organic coating system should be the minimum number used. If the existing paint or coating is sound, primers may be omitted from previously painted or coated surfaces, but all other coats recommended by the manufacturer should be applied to every surface, including previously painted or coated surfaces. This should be done even when the existing paint or other organic coating is sound, unless the paint or coating's manufacturer specifically recommends against it in a particular situation. When properly applied, the specified number of coats should produce a fully covered, workmanlike, presentable job. Each coat must be applied in heavy body, without improper thinning. When stains, dirt, or undercoats show through the final coat of paint or other opaque organic coating, the defects should be corrected and the surface covered with additional coats until the paint film is uniform in finish, color, appearance, and coverage. When defects show through a clear organic finish, the finish should be removed, the defects corrected, and the surface recoated.

Paint and other organic coatings should be applied at such a rate of coverage that the fully dried film's thickness for each coat will be not less than that recommended by the material's manufacturer. Special attention should be given to ensuring that edges, corners, crevices, welds, and exposed fasteners receive a dry-film thickness equivalent to that on the flat surfaces.

The first coat of paint or other organic coating material should be applied to prepared surfaces as soon as practicable after preparation, before surface deterioration can begin. Sufficient time should be allowed between successive coatings to permit proper drying. A minimum of twenty-four hours is required between applications of interior coats and forty-eight hours between exterior coats. Surfaces should not be recoated until the previous coat has dried until it feels firm, does not deform or feel sticky under moderate thumb pressure, and the application of another coat of paint or other coating will not cause lifting or loss of adhesion of the undercoat.

It is necessary to sand lightly between succeeding coats of most field-applied organic coatings. High-gloss paint should be sanded between coats using very fine grit sandpaper. The dust should be removed after each sanding to produce a smooth, even finish. Each coat of a paint or other organic coating system should be inspected and found satisfactory before the next coat is applied.

The edges of paint adjoining other materials or colors should be made sharp and clean, in straight lines and without overlapping.

Oily rags, oil, solvent-soaked waste, discarded paint and other organic coating materials, rubbish, and cans should be removed from buildings at the end of each work day. Precautions should be taken to avoid the danger of fire.

After paint and other organic coatings have been applied, damaged paint and other organic coatings should be examined, touched up and restored where damaged, and left in proper condition. Then glass and other spattered surfaces should be cleaned. Spattered paint and other organic coatings should be removed by scraping or other methods that will not scratch or otherwise damage finished surfaces.

Where to Get More Information

One of the best sources of information available today about metal finishes is the National Association of Architectural Metal Manufacturers' *Metal Finishes Manual for Architectural and Metal Products*. Anyone interested in designing, applying, or maintaining nonstructural metals or their finishes should have access to a copy of the latest edition.

The book in this series entitled *Repairing and Extending Finishes, Part II*, contains a more detailed discussion of field-applied paint than does this book. The discussions there also generally apply to shop-applied paint and other organic coatings. Because of book length limitations, all of the information in that book has not been repeated in this one. The bibliographical references there that are applicable to the painting and coating of metals, however, have been repeated in this book's Bibliography. The applicable listings under "Where to Get More Information" from that book have also been repeated here.

Many national and regional paint and coatings manufacturers offer quite good information about the systems they manufacture. As mentioned before, a reputable manufacturer may be the best source of advice about repairing a failed organic coating.

The books, periodicals, and other publications and standards listed in the following paragraphs are helpful when selecting or specifying paint and other organic coatings for use on new or existing metal surfaces. Some will also be helpful in determining how to maintain and repair such surfaces.

The following AIA Service Corporation's *Masterspec* Basic Version sections contain helpful information about the types of organic coatings addressed in this chapter. The author recommends that anyone involved in selecting paint or organic coating systems obtain and refer to a copy. The author assumes that *Masterspec*'s later editions will contain similar data.

- Section 09800, Special Coatings (February 1988). Includes discussions of epoxy, polyurethane, chlorinated rubber, heat-resistant enamel, black enamel, silicone-alkyd enamel, zinc rich, and aluminum-pigmented coatings.

- Section 09900, Painting (November 1988). Includes a discussion and a detailed listing of the national and regional paint and organic coatings manufacturers operating at the time. Regional manufacturers are broken down into those in the northeast region, southern states and gulf coasts, north central states, southwest region, and northwest states. Also includes a discussion of requirements for selecting paint and organic coating products and descriptions of their components that are more extensive than those in this chapter.

The following Copper Development Association publications contain information related to the subjects in this chapter:

- "Clear Film Coated Copper for Decorative Applications."
- *Copper, Brass, Bronze Design Handbook: Architectural Applications.*
- *Copper, Brass, Bronze Design Handbook: Sheet Copper Applications.*

- "Properties of Clear Organic Coatings on Copper and Copper Alloys."
- "How to Apply Statuary and Patina Finishes."
- "Clear Organic Finishes for Copper and Copper Alloys."
- "Take Care of the Metal in Your Building."

The following American Architectural Manufacturers Association publications contain valuable information for anyone interested in organic finishes on aluminum:

- "Voluntary Guide Specifications for Cleaning and Maintenance of Painted Aluminum Extrusions and Curtain Wall Panels." Describes recommended procedures for cleaning and maintaining painted aluminum after it has been installed.
- "Voluntary Performance Requirements and Test Procedures for Pigmented Organic Coatings on Extruded Aluminum."
- "Voluntary Specifications for High Performance Organic Coatings on Architectural Extrusions and Panels."

For excellent guidance in dealing with conditions where new paint is to be applied over an existing surface, see the editions listed in the Bibliography of the following:

- U.S. Department of the Navy, Guide Specifications Section 09910. "Painting of Buildings (Field Painting)."
- U.S. Department of the Army, CEGS-09910, "Painting, General."
- U.S. General Services Administration, PBS 3-0990, "Painting and Finishing, Renovation, Repair, and Improvement."

Later editions of the Navy and Army guides should be equally helpful. However, current editions of the GSA guides are based on *Masterspec* and may not be as specifically helpful for dealing with existing surfaces.

The U.S. General Services Administration's Specifications Unit's Federal Specifications applicable to organic coatings are sometimes out of date or not available, and the quality level they require is sometimes substandard. Unfortunately, they are often the only standard applicable. Some of the many Federal Specifications applicable to materials usable on metal are more often referred to than are others. The following list, which is by no means complete, contains some of the most frequently referred to specifications. Complying with the listed Federal Specification does not necessarily guarantee that a product will be satisfactory for a particular condition.

- Federal Specification TT-C-535. "Coating, Epoxy, Two-Component, For Interior Use On Metal, Wood, Wallboard, Painted Surfaces, Concrete and Masonry."

- Federal Specification TT-C-542. "Coating, Polyurethane, Oil-free, Moisture Curing."
- Federal Specification TT-E-489G. "Enamel, Alkyd, Gloss (For Exterior and Interior Surfaces)."
- Federal Specification TT-E-496. "Enamel: Heat Resisting (400 Deg F), Black."
- Federal Specification TT-E-505A. "Enamel, Odorless, Alkyd, Interior, High Gloss, White and Light Tints."
- Federal Specification TT-E-508C. "Enamel, Interior, Semigloss, Tints and White."
- Federal Specification TT-E-509B. "Enamel, Odorless, Alkyd, Interior, Semigloss, White and Tints."
- Federal Specification TT-E-545. "Enamel, Odorless, Alkyd, Interior-undercoat, Flat, Tints and White."
- Federal Specification TT-E-1593. "Enamel, Silicone Alkyd Copolymer, Gloss (For Exterior and Interior Use)."
- Federal Specification TT-L-190D. "Linseed Oil, Boiled (For Use in Organic Coatings)."
- Federal Specification TT-L-201A. "Linseed Oil, Heat Polymerized."
- Federal Specification TT-P-28E. "Paint, Aluminum, Heat Resisting (1,200 F.)."
- Federal Specification TT-P-30E. "Paint, Alkyd, Odorless, Interior, Flat White and Tints."
- Federal Specification TT-P-37D. "Paint, Alkyd Resin; Exterior Trim, Deep Colors."
- Federal Specification TT-P-47F. "Paint, Oil, Nonpenetrating-flat, Ready-mixed Tints And White (For Interior Use)."
- Federal Specification TT-P-81E. "Paint, Oil, Alkyd, Ready Mixed Exterior, Medium Shades."
- Federal Specification TT-P-86G. "Paint, Red-lead-base, Ready-mixed."
- Federal Specification TT-P-615D. "Primer Coating, Basic Lead Chromate, Ready Mixed."
- Federal Specification TT-P-636D. "Primer Coating, Alkyd, Wood And Ferrous Metal."
- Federal Specification TT-P-641G. "Primer Coating; Zinc Dust–Zinc Oxide (For Galvanized Surfaces)."
- Federal Specification TT-P-645A. "Primer, Paint, Zinc Chromate, Alkyd Type."
- Federal Specification TT-P-664C. "Primer Coating, Synthetic, Rust-inhibiting, Lacquer-resisting."

- Federal Specification TT-P-1511. "Paint, Latex-base, Gloss and Semi-gloss, Tints and White (For Interior Use)."
- Federal Specification TT-T-291F. "Thinner, Paint, Mineral Spirits, Regular And Odorless."

There are more than 600 ASTM standards related to paint and organic coating materials. Most of them, however, are testing requirements for chemicals that might be used in paints and organic coatings products and are, therefore, not too helpful in selecting new paints or organic coating materials or for determining the type of paint or organic coating in an existing building. The following list contains some of the ASTM standards that might be helpful.

- Standard D 16, "Standard Definitions of Terms Relating to Paint, Varnish, Lacquer, and Related Products."
- Standard D 1730, "Recommended Practices for Preparation of Aluminum and Aluminum-Alloy Surfaces for Painting."
- Standard D 1731, "Recommended Practices for Preparation of Hot-Dip Aluminum Surfaces for Painting."
- Standard D 2833, "Standard Index of Methods for Testing Architectural Paints and Coatings."
- Standard D 3276, "Standard Guide for Painting Inspectors (Metal Substrates)."
- Standard D 3927, "Standard Guide for State and Institutional Purchasing of Paint."

The following publications from the Steel Structures Painting Council should be available to everyone responsible for painting steel:

- The 1983 *Steel Structures Painting Manual, Vol. 1, Good Paint Practice, Second Edition.*
- The 1983 *Steel Structures Painting Manual, Vol. 2, Systems and Specifications, Second Edition.*

The 1982 American Society for Metals' *Metals Handbook: Ninth Edition, Volume V: Surface Cleaning, Finishing, and Coating* contains very detailed and excellent discussions of materials and methods for preparing for and applying organic coatings on metals. It also discusses the reasons for failures. It is an excellent source of data for anyone who is responsible for dealing with the finishes used on metals.

The June 1988 article "Direct to Rust Coatings" in *American Paint Contractor* contains a good description of the various materials available for painting rusted metals, and their application.

Timothy B. McDonald's 1987 article "Technical Tips: Coatings That Protect Against the Corrosion of Steel" contains the criteria for selecting organic coatings for steel, and other important recommendations regarding primers, curing, and what to do about already corroded surfaces.

The National Decorating Products Association's 1988 *Paint Problem Solver* is a definitive source specifically addressing paint problems and their solutions. It covers alligatoring, blistering, checking, cracking, flaking, excessive chalking, peeling from galvanized metal, metal doors, and garages, sagging, wrinkling, and peeling of latex top coats from previously painted hard, slick surfaces, and more. It also discusses application problems, including an applicator's not holding enough paint, brush marks, cratering, excessive shedding of bristles onto a painted surface, and excessive splatter from rollers. Discolorations, including fading, mildew, rusted fastener heads, and staining from flashings, are also covered. Various other problems are covered as well, such as lap marks, poor hiding, and uneven gloss. Everyone responsible for paint maintenance and painting over existing materials should have a copy of this work. It is also available from the Painting and Decorating Contractors of America (PDCA).

The following Painting and Decorating Contractors of America publications are of varying value, as noted, for those having to deal with the maintenance of paint or the painting of existing surfaces. Some critics have, rightly, pointed out that much of the data in these PDCA publications are also available from some major national paint manufacturers. However, there is a twofold advantage of having the PDCA data. First, it is a source of needed data that is always available, even when the selected manufacturer has not published similar recommendations, which is often the case. And second, it serves as a second opinion as compared with data in other sources.

- The 1975 edition of *Painting and Decorating Craftsman's Manual and Textbook, Fifth Edition.* Used in training craftspersons.
- The 1982 *Painting and Decorating Encyclopedia.* A good resource for designers, decorators, and craftspersons. Not specific on existing conditions.
- The 1986 *Architectural Specification Manual, Painting, Repainting, Wallcovering and Gypsum Wallboard Finishing, Third Edition.* Actually published by Specifications Services, Washington State Council of the PDCA in Kent, Washington. It should be on the shelf of every person responsible for designing, maintaining, and repairing painted surfaces. Includes an evaluation of finishing systems and discusses new and existing surface preparation, general information and finish schedules for interior and exterior paint finishes, and guide specifications. These specifications are not as definitive as others available and are not based on the current Construction Specifications Institute format. The parts of this book about repainting discuss how to handle existing

sound, slightly deteriorated, and severely deteriorated paint on metal (including preparation methods for repainting) for ferrous and galvanized metal and aluminum. It covers cleaning and removal of existing paint by hand, solvents, steam, power tools, burning off, chemicals, and sandblasting. It also includes removal methods for efflorescence and mildew, and acid-etching methods. Other subjects are methods for handling extractive bleeding from cedar, redwood, mahogany, and Douglas fir.

- The 1988 *The Master Painters Glossary*. An excellent glossary of paint-industry terminology that is well worth having.

Thomas R. Scharfe's article "New Metal Coating Technologies Enhance Design Opportunities" includes a list of coatings and some of their characteristics.

Forrest Wilson's 1984 book *Building Materials Evaluation Handbook* offers sound advice about cleaning metal surfaces.

The Construction Specifications Institute's 1988 "CSI Monograph Series, 07M411, Precoated Metal Building Panels" contains basic information about the coil-coating process and gives the sources of the information it contains.

C. R. Bennett's 1987 "Paints and Coatings: Getting Beneath the Surface" is as complete a general introduction to paints and coatings as one would expect to find in a magazine article. In a letter to the editor Mike Bauer takes issue with some of Mr. Bennett's statements, however, so the two pieces should be read together.

Larry Jones's 1984 article "Painting Galvanized Metal" is a good discussion of the problems associated with the subject, which is one of the more difficult surfaces to paint successfully. Some of his recommendations are not universally accepted, however.

Maurice R. Petersen's series of articles in 1984–1985, "Finishes on Metals: A View from the Field," discusses organic coatings used on metals, compares them, and includes recommendations about repairing coatings.

Martin E. Weaver's 1989 series on "Fighting Rust" offers an excellent discussion of the causes and cures for rusted iron and steel.

Bernard R. Appleman's 1989 article "Coatings for Steel Structures" discusses the environmental hazards associated with coatings, and the regulation of them.

In addition, the following articles and books may prove of value:

- Kenneth Abate's 1989 article series "Metal Coatings, Fighting the Elements with Superior Paint Systems."
- Able Banov's 1973 book *Paints and Coatings Handbook for Contractors, Architects, and Builders.*

- *Architectural Technology*'s 1986 article "Technical Tips: Paints and Coatings Primer."
- David R. Black's 1987 article "Dealing with Peeling Paint," Construction Specifications Institute's 1988 Monograph 07M411, "Precoated Metal Building Panels," and the 1988 *Specguide* 09900 "Painting."
- Caleb Hornbostel's 1978 book *Construction Materials, Types, Uses, and Applications.*
- Robert Lowes's May/June 1988 article "Abrasive Blasting" in *PWC Magazine.*
- Ambrose F. Moormann, Jr.'s 1982 article "Paint and the Prudent Specifier."
- Dave Mahowald's 1988 article "Specifying Paint Coatings for Harsh Environments."
- Charles R. Martens's 1974 book *Technology of Paints and Lacquers.*
- Harold B. Olin, John L. Schmidt, and Walter H. Lewis's 1983 edition of *Construction Principles, Materials, and Methods* is a source of data about the organic finishes addressed in this chapter. Unfortunately, there is little in their book that will help with repairing these finishes.
- The U.S. Department of the Army's 1980 *Painting: New Construction and Maintenance* (EM 1110-2-3400).
- The U.S. Department of the Army's 1981 *Corps of Engineers Guide Specifications, Military Construction*, CEGS-09910, "Painting, General."
- The U.S. Departments of the Army, the Navy, and the Air Force's 1969 *Technical Manual TM 5-618, Paints and Protective Coatings.*
- The U.S. Department of Commerce, National Bureau of Standards 1968 *Organic Coatings BSS 7.*
- The Guy E. Weismantel edited 1981 *Paint Handbook.*

CHAPTER

6

Design and Fabrication of Nonstructural Metal Items

The data in this and the next chapter are generally applicable to all nonstructural metal items (metal fabrications), as defined in this book. The discussion may or may not apply to structural metals or to the nonstructural metal items listed as being excluded from this book under ''What Nonstructural Metals Items Are and What They Do'' in Chapter 1.

Support systems for the metal fabrications discussed in this and Chapter 7 are covered in Chapter 2, the metals used to fabricate them in Chapter 3, the finishes for them in Chapters 4 and 5, and the installation of them (including means and methods of anchoring and fastening them in place), repair, and cleaning in Chapter 7.

Typical Nonstructural Metal Items

Some Types of Items

There are many types of metal fabrications to which the principles addressed in this and the next chapter apply, including those that follow.

Carpenters' Iron Work.   Carpenters usually furnish their own straight fasteners, such as nuts and bolts, screws, and nails. There are, however, always required certain bent or otherwise custom-fabricated bolts, plates, anchors, hangers, dowels, and other miscellaneous steel and iron shapes.

Anchors and Fasteners.   Such devices are used to support and fasten nonstructural metal items in place.

Loose Plates and Shapes.   Loose bearing and leveling plates for steel items that bear on masonry or concrete construction, and loose lintels, shelf angles, and beams with hung plates to support stone or masonry are usually considered nonstructural metal items.

Framing, Supports, and Reinforcements.   This category comprises the above- and below-ceiling support systems for equipment of all kinds, stationary and operating partitions, doors, opening frames, mall fronts, stage curtains, toilet partitions, and other items.

Partition Reinforcement for Door Frames.   Automatic doors and heavy doors, such as lead-lined ones, often require additional metal supports.

Ladders.   All kinds of fixed ladders, including straight ladders and ships' ladders, are considered nonstructural metal items. Ladders are used for access to roofs, elevator and other pits, and other otherwise inaccessible spaces.

Stairs.   Metal stairs of all kinds are considered metal fabrications. The different types include all-steel stairs, steel stairs with concrete or stone treads, and aluminum stairs. Stairs may be straight, of the scissors type, or spiral. Their metal treads may be of flat patterned plate, gratings, or other configurations.

Handrails and Railings.   There are many types of handrails and railings, some decorative but some simply utilitarian. Either type may be found either on the exterior or the interior of a building. Both types may be used in conjunction with steel or concrete stairs, steps, and ramps; along the tops of walls; around floor openings; as guard rails; as handicapped-assistance rails; as hospital corridor patient rails; and for many other uses. Some railings include operable gates.

Utilitarian handrails and railings are usually fabricated from steel pipe, tubing, bars, or other shapes. There are many possible configurations, from straight-steel-pipe wall railings to elaborate post-and-newel railings with balusters or pickets. Many center railings in stairs are simply steel pipes, but some have vertical bars, grilles, wire mesh, or other infill. Some have solid infill panels below their handrails. The term "utilitarian" does not necessarily imply that such railings must be ugly or even simple, but rather has to do with their use and location. Some utilitarian-type railings are used in highly visible locations, however, such as lobbies, where they might also be considered decorative.

The term "decorative" is usually applied to handrails and railings in highly visible settings where appearance is important. These rails and railings may be of wood supported by metal wall brackets; made from metal tubes, pipes, bars, and other shapes; be metal gratings or grilles supported by tubes, bars, or other shapes; be glass panels supported by metal bars, tubes, and other shapes; and many other possible configurations. The metals used may be aluminum, bronze, or other decorative metals. Some decorative railings are lighted by internal devices.

Fences.   Almost by definition, fences are not used on the interiors of buildings. Even so, they are a kind of metal fabrication so often associated with buildings that to fail to mention them would be derelict.

There are many types of fences, some decorative and some purely utilitarian. Most include some kind of gate (Fig. 6-1).

Decorative fences (Fig. 6-2) may also be utilitarian, of course, but the opposite is usually not true, as can be clearly seen from the chain-link and steel-bar fence in Figure 6-3. Chain-link fences may be of steel or aluminum. Steel components are often galvanized with aluminum being often clear anodized. Either may be coated with a plastic, such as polyvinyl chloride (PVC). Decorative metal or plastic strips are sometimes slipped into chain-link fencing to make them more decorative and block the view through them.

Many older fences were made of wrought iron (Fig. 6-4; see also Fig. 6-1). Newer decorative fences are more likely, however, to be made of mild steel or aluminum. Ferrous-metal fences are usually finished with paint or another opaque organic coating. Aluminum may be clear or color anodized, or be finished with paint or another opaque organic coating. Some decorative metal fences are finished with a plastic cladding.

Gratings.   Many types of gratings are used in the great range of installations seen on the interiors and exteriors of buildings. The metals usually used are aluminum and steel. Gratings may be of the straight or deformed bar type, expanded mesh, pressed plates, or castings. The components may be welded together, bolted or screwed together, fastened or locked together by tabs, or riveted together. Their top surfaces may be smooth, serrated, or contain abrasive strips or coatings to give them nonslip textures. They may be either fixed or removable and have hinged or lift-out access panels.

Figure 6-1 A slightly different type of operating gate.

Gratings are used as stair treads and landings, pit covers, floor opening covers, areaway covers, and to protect the base of trees, and in other locations.

Floor Plates.   Steel plates are used to cover floor openings, as stair treads and landings, and for other purposes. Plates may be totally flat or have a raised pattern. The most common such pattern is the diamond, but other designs are also used.

Bollards.   There are many different types of bollards. Some are proprietary items made from metal, but others are just galvanized steel pipe filled with concrete. Whatever their construction, they all have the same function: to prevent vehicular contact with a fixed object, such as a part of a building that might be damaged (Fig. 6-5).

Figure 6-2   A decorative fence.

Figure 6-3   A purely utilitarian fence.

Figure 6-4 The top of a wrought-iron fence.

Window-Cleaning Devices. Buildings that are too tall for their windows to be reached by ladder are often fitted with some type of device to lower a portable platform down the face of the building. These devices may be as simple as window-cleaning hooks or as complicated as fixed outriggers. Some may be quite complex outriggers that telescope and travel horizontally.

Miscellaneous Metal Fabrications. The category called "miscellaneous metal fabrications" is one that architects often use to include many minor metal fabrications. It includes such miscellaneous iron and steel items as beams, plates, flats, angles, tees, channels, and the like that are not otherwise specifically classified. Such items are usually fabricated from structural steel shapes and plates and steel bars. Among the specific items often

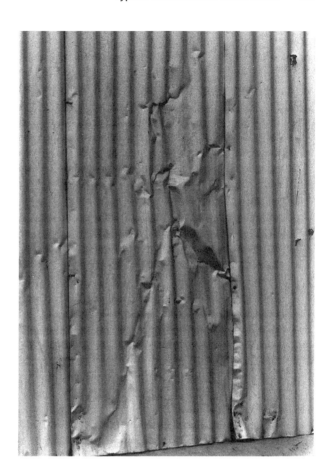

Figure 6-5 The damage in this photograph resulted from not protecting the building from vehicular contact.

included in this category are welded-steel angle frames for roof and floor openings, concrete-slab edge angles, clips for bracing masonry walls, elevator-sill support angles, overhead coiling door-jamb protectors and operator mounting plates, elevator hoist and shaft divider beams and counterweight divider beams, parking-area corner guards, locker-room bench supports, partition reinforcements for door frames, speaker supports, countertop support angles, elevator cylinder support channels and related elevator items, and metal grating supports.

Stair Nosings.   Metal stair nosings are used on concrete steps and stairs. Some have abrasive strips embedded in them, or they have surfaces configured to reduce slipping.

Fire-extinguisher Cabinets.   There are several types and styles available. They are usually of painted steel or aluminum but are also available as color-anodized aluminum, bronze, brass, and stainless steel decorative metals and glass-front units.

These cabinets may be designed and sized to contain only a fire extinguisher but may also be combination units housing fire hoses and other fire-fighting devices. The units may be fully recessed, partially recessed, or surface mounted.

Postage Accessories.   There are several types of postage accessories, ranging from individual mail-reception boxes to elaborate lockbox mail-receiver combinations. Such units may be of either the interior or exterior type. The individual boxes that make up the units are available in many sizes and combinations.

Units may be frontloading, meaning that the mail is placed in the boxes by removing their fronts, or backloading. The fronts on frontloading units are hinged. Backloading units do not always have backs but sometimes do. Some units are nothing more than pigeon holes, without either fronts or backs.

The materials used may be painted steel, painted terneplate, anodized aluminum, or bronze.

Floor Mats and Frames.   Floor mats may be made of metal but are more usually of rubber, carpet, vinyl, cocoa mat, or some other material, though some mats do have metal components or dividers. Frames are usually of stainless steel or anodized aluminum.

Signs and Directories.   Today many signs and directories are made of plastic. Some, however, are still produced in decorative metals, including stainless steel, anodized aluminum, and copper alloys.

Exterior Metal Screens.   Screens of many different types are used to conceal unattractive exterior equipment or parts of buildings. One example is a roof-top screen used to conceal fans, cooling towers, and other mechanical equipment.

Access Panels and Hatches.   Access panels, or doors, are used to gain access to mechanical, electrical, and plumbing equipment and devices. They are installed in concrete, masonry, plaster, gypsum board, ceramic tile, and other wall and ceiling surfaces. Both unrated and fire rated units are available. Most are constructed from sheet steel. Some are painted or given a baking-finish organic coating in the shop. Others are recessed to allow the same material used in an adjacent wall or ceiling to be inserted, to make the panel less noticeable. Some have key locks.

Floor Hatches.   Heavy plate–topped floor hatches are used to gain access to mechanical, electrical, plumbing, elevator, and other equipment. Some floor hatches provide openings to pass materials from one floor to another without sacrificing valuable floor space, as would happen if a permanent opening were provided. Floor hatches may be either single- or double-leaf. Most have checkered floor-plate covers.

Lockers.   There are many configurations for lockers, including single tier, multiple-person, and combination units of various types. Lockers are usually constructed of sheet steel with a baking finish, but steel-mesh lockers are also available. Various types of fronts are available, including solid, perforated, grilled, louvered, and steel mesh. The sides and backs may be solid, perforated, or of steel mesh.

Unit Kitchens.   Almost all unit kitchens are proprietary units containing all the necessary components, such as countertops, refrigerators, sinks, ranges, ovens, range hoods, and base and wall cabinets. They are manufactured mostly from carbon steel, stainless steel, and aluminum, and consequently most of the construction methods used in them are similar to those described in this chapter and Chapter 7. The same is true in fact for most residential appliances of all types.

Shelving.   There are several types of metal shelving. Sheet-metal shelving is used in utilitarian applications, such as storage rooms and commercial kitchens but also in more visible locations, such as in libraries. Shelving's finishes range from polished stainless steel to a baking finish on carbon steel. Some aluminum sheet-metal shelving is also used, but not as widely as that of steel.

Another type of shelving unit, used mostly in utilitarian applications but also in laboratories and laboratory storage rooms, consists of metal standards and either wood, pressed wood, plastic laminate, synthetic stone, or other nonmetallic shelves. The metal standards are usually of carbon steel with a baking finish, chromium-plated steel, or stainless steel.

Rod-type metal shelving is used in food-service areas, for clothes storage, in closets, and in many commercial sales areas. Stainless steel rod and vinyl-clad steel rod are two commonly used types, though aluminum-rod shelving is also available.

Most metal shelving units are proprietary products.

Expansion Joint Assemblies. Metal expansion joint assemblies are used to cover expansion joints at interior floors, walls, and ceilings that would otherwise be visible. They are usually not used in connection with acoustical ceilings. Expansion joint asemblies are also sometimes used in exterior locations where joints might otherwise be unsightly or create a hazard.

Most metal expansion joint assemblies are made of aluminum, but other metals, such as stainless steel and bronze, are sometimes used.

The configuration of expansion joint assemblies varies according to their location. A joint in a flat wall or floor may be covered by a flat plate mounted in an assembly with flanges and a compressible medium. The floor cover may be fluted for slip resistance. A joint in a corner may be covered with an L-shaped metal member. Joints that move in more than one direction may require elaborate assemblies to allow for movement in both directions.

Floor assemblies often contain sheet plastic waterstops to prevent water used to clean the floors from entering a ceiling space below. Exterior floor assemblies may have drainage systems to bleed off water that penetrates the assembly.

Many types of assemblies are available. Some are designed so that only the metal shows; others let plastic compressible material show. Some have slip-on or snap-on plastic covers.

Some metal joint covers snap in place, but others must be screwed on or bolted.

Rolling Counter Shutters. One method of closing an opening between spaces in which a pass-through function is necessary is to use counter shutters. These may be fire rated or unrated and constructed of steel or aluminum. The slats in them may be steel, stainless steel, aluminum, or even wood. The finish on the metal components may be a stainless steel finish, anodized aluminum, a baking finish, or paint. The operation of them may be manual or motorized. Manual operation may simply involve pushing them up by hand for small shutters or using a chain hoist assembly for larger shutters.

Shutters may be mounted under a lintel or on the inner face of a wall.

Flagpoles. There are many types and sizes of flagpoles. They may be manufactured of steel or aluminum and then either painted or finished with an integral finish, such as anodizing on aluminum. Flagpoles may be ground set, wall mounted, or mounted on top of another structure. They may be fixed in place or tilt down for access. The halyard, or flag-raising, mechanism may be manual or winch operated and either be externally mounted or contained within the pole. Their operation is usually manual, but they may also be motorized.

Commercial flagpoles usually have some sort of lightning-protection device built into the installation. For example, ground-set poles have steel spikes extending into the ground from the pole's base plate. Ground-set poles are installed by wedging them into a metal (usually galvanized steel) tube that is in turn set in concrete. The space between the pole and the tube is then filled with sand and the top sealed with a waterproofing compound.

Fireplaces and Flues. Many fireplaces are constructed from masonry using clay flue liners, but others are prefabricated metal units with metal flues. There are several types of prefabricated metal fireplaces on the market. Their features vary somewhat, but essentially they are all pre-engineered, heat-circulating metal fireplaces with pre-engineered metal chimneys. They come with every feature necessary for them to operate properly.

Even fireplaces constructed from masonry have some metal parts, including spittles, dampers, fronts, and chimney spark arrestors. Their fronts are often glass in brass frames, with brass screens and operating hardware.

Toilet and Bath Accessories. Items considered toilet and bath accessories include, but are not necessarily limited to, combination paper towel dispensers and waste receptacles, soap dispensers, paper towel dispensers, combination paper towel dispensers and soap dispensers, waste receptacles, feminine napkin-tampon vendors, feminine napkin disposals, toilet-seat cover dispensers, toilet tissue dispensers, toilet paper holders, wall urns, soap dishes, grab bars, framed mirrors with or without shelves, robe hooks, and mop and broom holders. Some items that are usually called toilet and bath accessories, including some of those in the preceding list, are clearly not actually so, but they are often included in this group anyway.

In general, the toilet accessories in a particular project should, as far as possible, be products from a single manufacturer for each type of unit and for all units exposed in the same area.

Stamped names or labels on the exposed faces of units are often prohibited.

Locks should be provided where they are either standard or optional with the units. All accessories in the same project should have the same keying whenever possible, except for coin-box locks, which should be keyed separately.

Most accessories are fabricated from either AISI Type 302/304 or 430 stainless steel with a No. 4 satin finish, or chrome-plated metal, brass, a zinc–brass alloy, or galvanized carbon-steel sheets complying with ASTM Standard A 366, with a baking finish or a paint coating.

The units may be partly or fully recessed or counter mounted.

Fabricated Metal Panels.   There are many types of fabricated metal panels available. They are made from various metals, but most are of carbon steel, stainless steel, galvanized steel, aluminized steel, or aluminum. Their finishes may also vary greatly, including coil-coated enamels, fluorocarbons, anodizing, silicone polyester, plastic coatings, other baking-finish organic coatings, or even field-applied paint.

Fabricated panels are available individually or as parts of wall systems. Some are just the sheet metal, but others consist of sheet metal applied to a backing, such as a foamboard insulation. Another type of panel is made of corrugated metal sheets, which are often cut to fit and then joined together in the field. The wall of the quonset hut in Figure 6-6 are an example of such construction.

Window Stools.   Both steel and aluminum are used for window stools in many commercial buildings. They are fabricated from sheet metals, usually without joints, where possible, and are reinforced on the concealed side when necessary. Most are finished with a baking finish or field paint.

Air Conditioning and Heating Enclosures.   Fin-tube radiators and similar heating devices are often not manufactured with their own enclosures and

Figure 6-6   The corrugated sheets on this quonset hut have seen better days.

grilles. Consequently, these items must be fabricated in a shop, from sheet metals and extrusions. Usually the cabinets are about 18- or 20-gage sheet steel with a baking finish or shop coat ready to receive a paint coat after application. The grilles are sometimes a part of the enclosure but are more usually a separate item. Separate grilles are often extruded-aluminum fins or gratings, or stamped plates with square holes. Sometimes these units are constructed of aluminum sheets or have extruded aluminum trim.

Wall Caps and End Closers. Low walls must often be capped with some material for the sake of appearance. Gypsum board and plaster low walls are often capped with formed sheet metal. The ends of such walls and partitions are often similarly capped. Such wall end and cap closers are usually fabricated from 18-gage sheet steel, then field painted.

Joints are a particular problem in such units that must be handled carefully if their appearance is to be satisfactory. Corners, for example, should be mitered and straight joints welded, with the welds ground smooth. The longest pieces possible should be used, to minimize joints. Exposed fasteners should be avoided.

Applicable Standards

Metal fabrications are subject to building code and other legal requirements just as are the other parts of buildings. These requirements often refer to industry standards as the applicable restrictions concerning certain kinds of metal fabrications. A building code might, for example, require that ladders comply with American National Standards Institute (ANSI) Standard A14.3 (Safety Code for Fixed Ladders). Unless more stringent requirements are written into the code or other applicable legal restriction, it is a good idea to require that metal fabrications meet applicable industry standards, expecially when the code is silent.

In addition to meeting the requirements of the building code and other governing laws, ordinances, and regulations, metal fabrications should comply with the applicable provisions of the following standards:

- Items constructed from structural-steel items and loose plates and shapes should comply with the American Institute of Steel Construction's *Specifications for the Design, Fabrication, and Erection of Structural Steel for Buildings.*
- Welding should comply with the applicable provisions of the American Welding Society's *Structural Welding Code.*
- Ladders should comply with ANSI Standard A14.3.
- Stairs and railings should comply with the National Fire Protection Association's *NFPA 101 code,* ANSI Standard A58.1, and applicable Na-

tional Association of Architectural Metal Manufacturers (NAAMM) recommendations.

- Gratings should comply with the NAAMM's *Metal Bar Grating Manual* or *Heavy Duty Metal Bar Grating Manual.*
- Fabrications containing sheet-metal components should conform to the Sheet Metal and Air Conditioning Contractors National Association's recommendations for fabrication and construction details, where applicable.
- Chain-link fencing should comply with the recommendations of the Chain Link Fence Manufacturers Institute's *Product Manual* as well as Federal Specifications RR-F-191/GEN, RR-F-191/1, RR-F-191/2B, RR-F-191/3, RR-F-191/4, and ASTM Standards A 120, A 123, A 153, A 392, A 491, B 221, and F 668 (when the fencing is coated with plastic).

Materials

This section contains a listing of the metal materials, fasteners, and accessories generally used to produce metal fabrications. Refer to each item listed in "Typical Nonstructural Metal Items" earlier in this chapter for the particular material that might be used for that type of fabrication.

Metal Products

Metal used to fabricate metal fabrications that will be exposed to view should be flat, smooth, and free from surface blemishes. Visible surfaces should not exhibit pitting, seam or roller marks, rolled trade names, roughness, oil canning, stains, discolorations, or other imperfections.

The following list contains many of the metal products one would expect to find in metal fabrications. Accompanying each item is the standard with which it is usually expected to comply and, in some cases, additional information related to the item and its selection for use in metal fabrications.

- Steel plates, shapes, and bars: ASTM Standard A 36.
- Mild steel plates, shapes, and bars: ASTM Standard A 283. Several grades are available. The proper one is usually selected by the fabricator.
- Steel tubing: cold formed, in accordance with ASTM Standard A 500, or hot rolled, in accordance with ASTM Standard A 501.
- Hot-rolled carbon-steel bars and bar-size shapes: ASTM Standard A 575. The grade used is usually selected by the fabricator.

■ Cold-finished steel bars: ASTM Standard A 108. The grade used is generally selected by the fabricator.

■ Structural cold-rolled steel sheet: ASTM Standard A 570.

■ Structural steel sheet: hot rolled, in accordance with ASTM Standard A 570, or cold rolled, in accordance with ASTM Standard A 611, Class 1. The grade should be selected to accommodate the design loading.

■ Galvanized structural-steel sheet: ASTM Standard A 446. The grade should be selected to accommodate the design loading. The coating designation generally used is G90.

■ Galvanized carbon-steel sheet: ASTM Standards A 525 and A 526, commercial quality, with coating designation G90 and mill phosphatized in accordance with ASTM Standard D 2092, Method A.

■ Zinc coating on steel mounting devices: hot-dip galvanized after fabrication, ASTM Standard A 386.

■ Cold-rolled carbon-steel sheet: ASTM Standard A 366, Class 1. It is usually matte finished.

■ Hot-rolled carbon-steel sheets and strips: ASTM Standard A 568 and ASTM Standard A 569. It should be pickled and oiled.

■ Steel pipe: ASTM Standard A 53. The type and grade is usually selected by the fabricator, but must be appropriate for the design loading. Pipe may be either black finish or galvanized. Most pipe used in fabrications is standard weight (Schedule 40), Type E or S, Grade B.

■ Gray-iron castings: ASTM Standard A 48, Class 30.

■ Malleable-iron castings: ASTM Standard A 47. The grade is usually selected by the fabricator.

■ Ductile iron: ASTM Standard A 536.

■ Cast steel: ASTM Standard A 27.

■ Stainless steel: The most commonly used type is probably American Iron and Steel Institute (AISI) Type 302/304, which is the common 18/8 material, but other types, such as AISI Type 430, are also used. Probably the most common finish used on stainless steel used in metal fabrications is the No. 4 satin finish, but any of the stainless steel finishes discussed in Chapter 4 can be used.

■ Steel floor plate: Federal Specification QQ-F-461C, Class 1. Most have a raised diagonal pattern, but other designs are also used. Such panels may also be flat, with no pattern.

■ Steel bar grating: ASTM Standard A 569 or ASTM Standard A 36.

■ Wire mesh: There are many available meshes, for example Woven Wire Products Association's No. 6 galvanized steel wire with a two-inch square mesh securely locked into a 3/4-inch-deep galvanized-steel channel frame. The frame's corners are usually mitered and welded.

- Metal lath: Federal Specification QQ-L-101, copper-alloy steel, galvanized, 3.4 pounds per square yard, standard metal lath.
- Aluminum extrusions: ASTM Standard B 221. Many common extrusions have properties of alloy 6063-T52, but other alloys, such as 6061, and other tempers are also frequently used.
- Aluminum sheet and plate: ASTM Standard B 209. Common alloys are 3003 and 5005, but others are also used. The temper should be selected as required for the forming involved and the finish to be applied.
- Bronze: There are many alloys of bronze. Some of them, such as extrusions, are made specifically for a particular product. They should be selected for compatibility with each other and their ability to produce the desired finish. Refer to Chapter 3 under "Copper Alloys" for a detailed discussion. Any of the bronze alloys discussed there and in the reference materials listed there can be used in metal fabrications.
- Brass: What has just been said about bronze also applies to brass.
- Chrome-plated metal: A common chrome-plated metal used in making toilet accessories is composed of nickel and chromium electrodeposited on metal in accordance with ASTM Standard B 456.
- Zinc–brass alloy: ASTM B 86. This material is commonly used to manufacture toilet accessories.

Miscellaneous Materials

Bituminous Paint. The Steel Structures Painting Council's Paint 12, cold-applied asphalt mastic, is often used where bituminous paint is required.

Mirrors. Toilet room and similar mirrors are usually required to comply with Commercial Standard CS 27.

Welding and Brazing Filler Metals, Solder, and Associated Materials. The types and alloys used should be those recommended by the producer of the metal to be welded and be as required for color match, strength, and compatibility in the fabricated items.

Mechanical Fasteners. Many types of fasteners are used to attach the various components of metal fabrications to each other in the shop. They include but are not limited to a great range of bolts, screws, rivets, pop rivets, explosive rivets, speed nuts, and quick-release fasteners. The proper types and sizes of these devices should also be supplied with the fabrications for use in field assembly, where this is necessary (see Chapter 7). For each bolt there should be the proper matching nuts and washers.

Generally, fasteners in metal fabrications should be of the same material as that being joined. When using the same material is not practicable or two different materials are to be joined, the fasteners should be of a material that will not cause galvanic—or any other type—of corrosion to occur in any part of the assembly. When exposed fasteners are necessary, they should be of the same color and finish as the metal fastened. For instance, fasteners of aluminum, stainless steel, or some other material compatible with aluminum should be used to fasten aluminum, and copper-alloy fasteners should be used to fasten copper alloys. Fasteners used in color-anodized aluminum should match the aluminum's finish. Similarly, fasteners used in copper alloys should be finished to match the color of the finish. Painted fasteners should not be used in color-anodized aluminum or copper-alloy surfaces—they will almost never match the metal's color and will be immediately and objectionably apparent.

Stainless steel fasteners are often used in stainless steel, chromium-plated, and aluminum items. Galvanized-steel fasteners are sometimes used in steel fabrications that will be exposed to exterior or severe interior environments, but in such instances corrosion-resistant fasteners are a better solution.

A partial listing of the types of fasteners and related devices generally used in metal fabrications follows. After each item is the standard with which that item should comply.

- Bolts and nuts: ASTM Standard A 307, Grade A.
- Lag bolts: Federal Specification FF-B-561.
- Machine screws: Federal Specification FF-S-92.
- Wood screws: Federal Specification FF-S-111.
- Plain washers, round, carbon steel: Federal Specification FF-W-92.
- Toggle bolts: Federal Specification FF-B-588.
- Lock washers, helical-spring-type carbon steel: Federal Specification FF-W-84.

Sleeves. Sleeves associated with metal fabrications are usually galvanized-steel pipes or tubes when they are to be built into adjacent construction, but other metals are also used. Small sleeves may be zinc. Exposed sleeves may be made from the same material as the metal to be inserted and have the same finish. Sleeves are usually furnished to the installer for field installation (see Chapter 7).

Sound-Deadening Material. Sound deadening is important in some types of metal fabrications, especially those made from sheet metals where impacts are normal. Usually, a heavy-bodied, fire-resistive coating is used where sound deadening is appropriate. Such coating materials are compounded for permanent, nonflaking adhesion to metal in a 1/8-inch thick coating. Other types of sound-deadening material may also be used.

Insulation. There are many types of insulation used in metal fabrications.

In fabrications other than panels that fill openings in the exterior shell of a building, the insulation is often similar to that used in exterior walls. For example, one type is semirigid, aluminum-foil faced, mineral-fiber felt, safing insulation, which is made for use as a fire stop. It is noncombustible and noncorrosive to aluminum and steel. The material should conform with the requirements of Federal Specification HH-I-521F, Type I or Federal Specification HH-I-558B, Form A, Classes 1 and 2. It should have the following fire-hazard classification ratings, in accordance with the tests in ASTM Standard E 84: a flame spread of 15 or less, a fuel-contributed rating of zero, and a smoke-developed rating of zero. It should be fastened in place using the type of fastener or adhesive recommended by the insulation's manufacturer.

Another common type of insulation used in metal fabrications other than panels is glass fiber, coated duct liner board of the same type used in heating and air-conditioning systems.

There are many types of insulation used in metal-faced wall panels. They include foam boards, such as polyisocyanurate and polyurethane foams, high-density glass-fiber boards, extruded and molded polystyrene, and others.

Sealants. Each fabricator has its own favorite type of metal-to-metal sealant. All such sealants, however, should be of a type that is permanently elastic. Oil-drying sealants should not be used.

Finishes

Chapters 4 and 5 contain detailed information about the finishes generally applied on the metals used in metal fabrications.

Ferrous Metals. Most metals used in nonstructural metal fabrications are ferrous and require some sort of applied finish. The exceptions are those with decorative metallic finishes, such as chromium plating, and most stainless steels. Some steel alloys that are specifically formulated to form a protective corrosive layer are not usually given an additional protective coating. Such steels are seldom used, however, in nonstructural applications.

Many, and perhaps most, of the items that the industry usually calls "metal fabrications" or "miscellaneous metals" are fabricated from ungalvanized ferrous metal. Then they are either factory or field painted, or finished in the shop or factory with a baked-on organic coating. Such fabrications that are to be field painted are usually first given a prime coat of paint in the shop. Some fabrications of this type are made from galvanized metals or are galvanized after fabrication. These are not usually given a primer coat of paint in the shop, but some are.

Most shop coats consist of a single coat. Two coats are often applied, however, to portions of metal fabrications that will be inaccessible after assembly or erection. All shop coats of paint are usually omitted from portions of metal fabrications that are to be embedded in concrete, stone, or masonry, from surfaces that will be field welded, and from galvanized surfaces.

Ferrous-metal anchors and miscellaneous items to be built into masonry, stone, or concrete are discussed in Chapter 7.

Sheet materials emerge from a rolling mill in long sheets that are then wrapped around a core in coils. The metal is then often given a metallic (anodizing, for example) or organic coating before it is cut from the coils. The usual process is to unroll a coil at one end of a coating line, pass the sheet through the coating mechanism, and recoil it at the other end of the line. When a baking finish is used, a baking chamber is imposed into the line. After this cycle, the coated coil is cut into proper lengths for its intended use. Refer to Chapter 5 for additional information on coil coatings.

Some metal fabrications contain porcelain-enamel coated or laminated coated ferrous metals (see Chapter 5).

Nonferrous Metals.   The finishes that may be used on nonferrous metals are addressed in Chapters 4 and 5.

Aluminum that is to remain concealed in metal fabrications is usually left with its mill finish. Exposed aluminum is almost always either anodized or coated with a fluoropolymer or other organic coating. Where appearance is not critical or the cost of colored finishes is too high for a project's budget to absorb, exposed aluminum is often given a clear anodized coating and finished with a methacrylate lacquer.

Much aluminum is given a colored anodic coating. Where the surface will be exposed to abuse or weathering, the usual coating is architectural class I (see Chapter 4). Hard-coat anodizing, which is a thicker than usual architectural class I coating, is often used in severe-use locations such as on entrance doors. Aluminum with less exposure may be given an architectural class II coating. Clear anodized aluminum is often protected by a methacrylate lacquer coating. Color-anodized aluminum should not be given a methacrylate lacquer finish, however, because it tends to not fully adhere, which leaves white spots on the finish, spoiling its appearance. When color-anodized aluminum must be protected, a strippable coating should be used.

A modern organic coating material that has enjoyed great popularity in recent years is thermo-cured fluorocarbon polymer. It is initially expensive, but it is durable, long lasting, and will maintain an acceptable appearance for many years. The better fluorocarbon polymer coatings in use at this writing (1989) consisted of an inhibitive thermo-cured primer of at least 2.0 mils dry thickness, with a thermo-cured fluorocarbon polymer top coating containing 70 percent Kynar 500 resin or its equivalent. The top coating should be applied at the rate necessary to form a coating of at least 1.0 mils in additional dry-film thickness. Many colors are available.

Other organic coatings may also be used on aluminum. They may be applied either to completed items or on the coil (see Chapter 5).

Any of the copper-alloy finishes mentioned in Chapter 4 may be used in metal fabrications. The mechanical finishes and the colors of the various elements in a metal fabrication made using copper alloys should match, unless a mismatch is specifically required by the design. Concealed copper-alloy surfaces that will be in contact with concrete, masonry, stone, or dissimilar metals should be coated with a heavy coat of bituminous paint to prevent corrosion. Such coatings should not extend onto exposed surfaces.

Copper alloys that will be exposed are often given a clear organic coating, as discussed in Chapter 5. One type often used is a 1.0-mil thick or thicker coating of the air-dried acrylic material called Incralac developed by the International Copper Research Corporation.

Design and Control

The procedures to be followed in designing a utilitarian metal fabrication and controlling its production and intallation so that the desired result is achieved will differ slightly, depending on whether the fabrication is part of new construction or is to be installed in an existing space. For purposes of discussion the term "new building" includes also additions to existing buildings, since all the work within an addition is actually new.

Design

When a fabrication is part of a new building, the normal procedure is for an architect, engineer, or other designer to produce design drawings showing the general configuration of the fabrication and then write specifications to describe the materials and methods to be used in producing the fabrication. Shop fabrication drawings are then made by the fabricator of the metal fabrication, based on the design drawing. The shop drawings describe the fabrication in greater detail than do the design drawings and include all the information the fabricator needs to actually produce the fabrication. The shop drawings are then submitted to the designer and are, when they seem to be correct, approved by the designer. It is usually a good idea— and is often a requirement of the contract between the owner and the building contractor—that measurements be taken in the field before metal fabrications are actually manufactured, and often even before shop drawings are made, to ensure that the actual conditions in the field match those shown on the original design drawings. The fabrications are then produced, based on the approved shop drawings and field measurements.

When a fabrication is to be installed in an existing space a design drawing is made, based on measurements taken in the field. When an architect, engineer, or other designer is involved, the process is similar to that just described for fabrications in new construction. When there is no designer involved, however, the shop drawing and the design drawing may both be produced by the fabricator—and may even be the same drawing. All drawings should be prepared from measurements made at the actual location where the fabrication will be installed. If the space there is to be modified in any way, measurements should again be taken after those modifications have been made.

Sometimes the time schedule of a project will require delivery and installation of a metal fabrication so early that delaying the manufacturer of the fabrication until other work has been completed in the area where the fabrication will be installed is not possible. Then the fabrication must be designed in such a way that it can be adjusted or trimmed to fit into a space slightly different in size or shape from that on the original drawings.

Control

In order to ensure that a fabrication actually matches the designer's intention and satisfies the owner's need, controls are instituted at every stage of the process. The first control exercised is the submission and review of the shop drawings mentioned. In addition, the fabricator and the installer of the fabrication are usually required to submit further detailed information to the owner.

The procedures for handling these submittals differ slightly, depending on several factors. All submittals must ultimately go to and be approved by the owner, who is, after all, paying for the fabrication. In a normal construction project, when an architect, engineer, or other design professional and a general contractor are involved submittals are usually reviewed first by the general contractor, and then the designer, before they are submitted to the owner. When there is no design professional involved, the general contractor should review the submittals before they are forwarded to the owner for review. They are usually sent directly to the owner without professional review only when the owner is dealing directly with the fabricator and there is no designer or general contractor.

Shop and Setting Drawings, and Manufacturers' Literature. The submittals usually required include the shop drawings mentioned. The classification "shop drawings" includes standard drawings and descriptions the manufacturer produces to describe the fabrication and drawings made specifically for the project at hand. Together they should show the methods of fabrication, assembly, and installation of the metal fabrications proposed by the fabricator and installer. The fabricator and installer may be the same entity, or different organizations altogether.

The shop drawings should include plans and elevations drawn at not less than a 1 inch to 1 foot scale, with details of sections and connections drawn at not less than 3 inch to 1 foot scale. The drawings should show materials, methods, joinery, fasteners, anchorage, and accessory items, method of construction and assembly, details of welded, bolted, and riveted connections, the kind and gage of metals, and the finish of the various parts. Electrical and plumbing roughing in drawings should also be included when applicable. When the materials used or the completed fabrications are subject to code or design criteria related to their ability to carry specific loads, which is often the case for some kinds of fabrications (equipment supports and stairs, for example), the shop drawings should also include structural computations, materials' properties, and other information needed for structural analysis.

To ensure the proper location and installation of each item, there should be furnished for each fabrication setting drawings and diagrams, templets, instructions, and directions for the installation of sleeves as well as anchorages, such as concrete inserts and anchor bolts, and miscellaneous items having integral anchors to be embedded in concrete or masonry.

The submittals should also include the manufacturer's, fabricator's, and finisher's brochures, specifications, anchoring details, and installation instructions. These should be included not only for products used in metal fabrications, including paint products and grout, but also for completed items such as toilet and other accessories.

The shop drawings and descriptive literature should together fully describe the fabrication and its finish and installation, including materials, fabrication, dimensions, finishes, hardware, and installation details, including mounting heights.

Cutting and Drilling Drawings.   When a new fabrication is to be installed in an existing space, it is usually a good idea to require not only shop and setting drawings but drawings also of the portions of the existing building where the metal fabrication is to be installed. These drawings should show existing conditions and the size and location of openings to be drilled or cut in existing construction. They should be reviewed by someone knowledgeable about structural stability and design, to ensure that no essential structural element is compromised by work related to the metal fabrication. Such drawings will often be required by local building authorities, who may require an architect's or engineer's signature or seal before accepting them. No cutting or drilling of structural elements or walls should be started until these drawings have been approved by everyone who should see them.

Similarly, it is sometimes necessary to require cutting and drilling drawings when an existing fabrication is to be repaired, if doing so will require cutting or drilling into existing construction.

Samples.   Samples of each metal material and finish required should be submitted to the owner and approved before any metal fabrication is installed. The samples should be large enough to display adequately the characteristics of the materials that will be used to produce the fabrication and the finishes to be applied. Large samples (8 inches square, for example) are better than small strips or squares, because they are more likely to show the actual appearance of the material. Samples of extruded shapes and sheet materials should be 12 inches or so in length.

Sometimes samples are required in the shape of the elements that will be used in the fabrication, but usually samples are not required for standard shapes such as angles, channels, and the like. A decision about which samples to require should be based on the complexity of a fabrication and whether it is a more or less standard product. Fewer samples might be requested for a vertical ladder, for example, than for a spiral stair. When there are many possibilities for a configuration, as in the case of gratings or toilet accessories, a sample of the actual product proposed for use may be required, to verify that the item to be furnished is actually the same as the one intended by the design. Sometimes the owner will need to see a sample of the manufactured components to be used in a fabrication, a section of the fabricated item, or even a completed item.

In some cases the samples submitted will later be installed in the building. This is frequently true for hardware and accessory items. Arrangements for such use are usually made before the samples are originally sent to the owner, and usually only items in like-new condition are permitted to be installed.

The metal in every sample should be of the same alloy, thickness, and finish as that to be used in the final products. Where normal variations in color and texture are to be expected, as is true with color-anodized aluminum and copper alloys, the samples should include as many examples as are necessary to show the full range of variations. Include 6-inch-long samples of linear shapes, including each extrusion, 6-inch-square samples of plates, and full-size samples of pipe, at least 6 inches long.

Color Charts.   In order for an owner to make a final decision on the actual colors of metal fabrications to be used, samples or color charts are usually submitted by the fabricator, through the contractor. These should show the complete range of the manufacturer's or finisher's available colors. Where an existing item is to be repaired or its colors matched, the colors submitted should include the matching colors and finishes.

Accessories Schedule.   Metal fabrications producers are often required to submit to the building owner a list of accessories that will be supplied with the fabrication. Sometimes many different accessories are available, but not all of them may be needed. Conversely, some accessories may be required that are not standard with the unit. A schedule helps to ensure that all that are needed will be included.

Construction and Demolition Schedules.   A construction-progress schedule should be developed for every construction project, whether the work is all new or is part of a remodeling, restoration, or renovation project. The schedule should show each significant stage of the work, with a starting and completion date for each part. Additional schedules showing the amounts of each scheduled payment are also sometimes required. Schedules are often broken down by discipline and then again within each discipline. The electrical work of a project may occupy a separate part of the schedule, for example.

There are many ways to develop and break down a construction-progress schedule. Two forms of schedule often used are the bar-chart form, used for most small projects and many large ones, and the critical-path form, which is usually used only for large, complicated projects. Bar charts simply plot each separate activity along a time line. The length of the bar representing each activity indicates how much time it will take. The bars are separated, with each activity on a separate line. All of the several critical path methods in general use are more detailed than a bar chart, and thus more complex. They require the use of a computer. Whereas bar charts simply plot each activity separately, critical-path charts show the interrelations between the various tasks that must be performed throughout a construction project.

It is often useful to require that the organization that will do the work prepare and submit a proposed schedule specifically showing the demolition phase of the work that must be done before a construction progress schedule is finalized. This is especially so if the portion of the building where the new or remodeled metal fabrication will be located or portions of the building that will be used as access ways are to remain occupied during the construction work. The demolition schedule should show when each part of the work will be started and completed and in what ways the work will interfere with the owners' or occupants' use of the building during the construction period. After its approval by the owner, the demolition schedule should be incorporated into the construction progress schedule.

Narrative Descriptions.   When existing construction is involved, it is often useful for the owner to require submittal of narrative descriptions detailing specifically the materials and methods proposed for use in doing the work.

When existing metal fabrications are to be cleaned or refinished, the organization that will do this work should be required to submit for the owner's approval a fully detailed description of the materials and methods to be used. The owner should have this proposal reviewed by someone knowledgeable about such matters before granting permission. This submittal should cover all pertinent data about the materials and methods proposed and include the cleaning material manufacturers' brochures, as well as manufacturers' and industry associations' recommendations concerning the materials and methods proposed. The proposal should also include materials and methods proposed for use if the originally proposed materials and methods do not satisfactorily clean or refinish the fabrication. For example, it should describe the materials or methods to use if mild soap, mineral spirits, and water are not sufficient to clean existing aluminum.

Sometimes it is prudent for an owner to request submittal of descriptions of the materials and methods proposed for use in restoration and repair of an existing metal fabrication. The submittals should include the manufacturer's published data fully describing each material and product proposed for use, with a listing of applicable industry standards. There should be shop drawings showing details of each condition expected to be encountered, including but not necessarily limited to installation and anchoring details and the relationship to other construction of each material and item requiring installation or reinstallation as each condition is encountered. A detailed narrative description should explain the methods to be used in making repairs, installing new items, reinstalling removed items, and protecting existing materials, equipment, accessories, and finishes that are to be left in place while restoration or repairs are in progress. Such data, shop drawings, and descriptions should be submitted even when the materials and methods to be used are indicated in design drawings and specifications.

Fabrication

The methods used to produce a metal fabrication will probably not vary simply if the fabrication is to be installed in an existing building instead of new construction, a category which includes additions to existing buildings.

General Requirements

Metal fabrications should be produced from the type, size, thickness, and gage of metal and with the finish, dimensions, and details shown on the design drawings or otherwise approved by the owner. Metal fabrications should be prepared in accordance with the manufacturer's latest published instructions and the approved shop drawings. Metal fabrications and their components and finishes should comply with applicable codes and be of the strength and durability necessary for the intended use. Their exposed surfaces should be smooth, flat, and free of imperfections. Their components should be formed in the maximum lengths and widths practicable with joints kept to a minimum. Cut edges should be concealed. The dimensions should be those shown on the approved shop drawings. Only previously proven details of fabrication and support should be used. All ferrous items should either be shop painted or galvanized after fabrication, as discussed in Chapters 4 and 5. Nonferrous items may, but may not be painted. Refer to Chapters 4 and 5 for a discussion of nonferrous metal finishes.

Exposed work should be formed true to line and level, with accurate angles and surfaces and straight, sharp edges. Except in rare cases, exposed edges should be eased to a radius of approximately 1/32 inch. Bent-metal corners should be formed to the smallest radius possible without causing grain separation or otherwise impairing the work. Joints should be concealed when possible.

Metalwork should be formed to the required shapes and sizes, with true curves, lines, and angles. The necessary rebates, lugs, and brackets should be provided so that the parts of a fabrication can be assembled. Metal fabrications should be cut, reinforced, and drilled as necessary to receive their finish hardware and related items.

Fabrications should be designed so that their components allow for expansion and contraction within a minimum ambient temperature range of 100 degrees Fahrenheit without buckling, excessive opening of joints, or overstressing of welds or fasteners.

Castings should be sound and free from warp, holes, and other defects that impair their strength or appearance. They should have joints where they will be the least conspicuous. Exposed surfaces should have a smooth finish and sharp, well-defined lines and arrises. Where machined joints are necessary, they should be milled to a close fit. The necessary rabbets, lugs, and brackets should be provided so that the work can be assembled in a neat and substantial manner.

Fabrications should be preassembled in the shop as much as possible, to minimize field splicing and assembly. Sometimes fabrications must be disassembled, to permit moving them to the installation site, especially when they are to be installed in an existing space. Whenever possible, however, units should be disassembled only when necessary for shipping and handling. Disassembled units should be clearly marked for ease in reassembly and for coordinated installation.

Matching trim should be furnished where necessary to complete the installation.

Joints

The component parts of metal fabrications should fit together accurately. Where exposed, components should meet at tight, hairline joints. Connections should be flush and smooth, with corner joints coped or mitered. Pipes and tubes should be joined using internal sleeves.

Joints that will be exposed to weather should be fabricated so that they will exclude water. Weep holes should be installed where water may accumulate.

Metallurgical Joining of Components.   There are essentially three different metallurgical methods of joining metals: welding, brazing, and soldering.

Before any type of metallurgical joining can be effectively accomplished, the metals involved must be properly cleaned and treated. Every bit of dust, dirt, oil, grease, scale, and other contaminants must be removed—even fingerprints can prevent proper joining. Oils and grease should be removed, using an appropriate commercial solvent. Where a flux is required, the surface must be wiped immediately with the proper cleaning agent before applying the flux, following the manufacturer's instructions.

Welding.   Simply put, welding consists of heating the metal on each side of a joint until it melts and fuses. Sometimes a filler is used between the two sides of the joint, and sometimes the metals are forced together under pressure.

Virtually all metals can be welded to themselves. A few different metals can also be welded to a few other metals, but most welding is done to join two pieces of the same metal. Many types of steel can easily be welded, either in the shop or the field. The lower the carbon content of the steel, the easier it is to weld. All stainless steel can be welded, but some types are more difficult to weld than others, austenitic stainless steel being the easiest. Caution must be exercised when welding stainless steel, however, for several reasons. For one thing, it is difficult to weld stainless steel without distorting the metal. Also, welding will often reduce or destroy its corrosion resistance. Successfully welding stainless steel in the field is difficult, because of the controls necessary to do it properly.

Aluminum can be welded fairly easily in the shop, but the methods required make the field welding of aluminum difficult.

Copper alloys are seldom welded, because the process tends to distort the metal significantly and change its color. Color matching a welded area to agree with adjacent areas is very difficult. Silicone bronzes can be successfully welded, however, so the process is used occasionally.

Thin metals are difficult to weld, because they tend to distort when melted. Welding a thin metal to a thick one is particularly difficult.

There are several processes used to weld metal. Three of the most widely used processes are gas, arc, and resistance welding. Arc welding is also called fusion welding.

In gas welding, the metal is melted by a burning gas. The gas used to weld steel is usually a combination of oxygen and acetylene, hence the term "oxyacetylene welding." In welding aluminum, hydrogen is sometimes substituted for the acetylene. Gas welding is seldom used in steel joints that must carry loads. It is generally limited to use in thin sections, in any metal. For example, aluminum thicker than 1 inch is seldom welded with the gas method, because the metal dissipates heat so rapidly that it is difficult to maintain the proper temperature.

The heat for arc welding is produced by passing an electric current between the metal and an external electrode. An important characteristic of electric arc welding is that the oxygen and nitrogen in air have a detrimental effect on the melted metal at the joint being welded. The solution is to shield the joint by enclosing the weld in an inert gas or covering it with a fusible granular material. This may be accomplished in several ways, including shielded-metal arc welding, which includes various methods: manual metal-arc welding and coated-stick electrode welding, gas tungsten-arc welding, gas shielded-metal arc welding, and submerged-arc welding. There are also other fusion welding methods used, including plasma-arc welding, electron-beam welding, and laser welding.

The heat for resistance welding is also produced by the flow of an electric current, but in this case the current passes from the metal on one side of the joint to the metal on the other side.

Resistance welding is often used on aluminum because it does not affect the temper of the metal. Flash welding, a resistance-welding technique, is often used to join mitered sections of aluminum.

Spot welding is a resistance-welding technique widely used with steel, stainless steel, and aluminum. In this method parts are joined by a series of small welds with spaces between them. Seam welding is a version of spot welding in which the spots are placed close enough together so that they overlap. Spot and seam welding are best used for multiple runs where the efficiency of the systems used to make them can best be taken advantage of.

Projection welding is another resistance-welding technique, where the weld points are predetermined by punching dimples in the metal at preset intervals. The welds then take place only where the projecting parts touch.

Other resistance-welding techniques include high-frequency resistance welding, percussion welding, and the commonly used method, butt welding. The two types of butt welding—flash welding and upset welding—are both shop processes.

There are other welding processes in use, including friction welding and electroslag welding. Both of these processes can take place only in the shop. In addition, special procedures are necessary when welding clad materials and different metals to each other.

Most of the time, welded corners and seams in metal fabrications should be welded continuously, not spot welded, so that they do not tend to distort along the seam. Welding should comply with the applicable recommendations of the American Welding Society. Welds should be made behind finished surfaces, without distortion or discoloration of the exposed side. Exposed welded joints should be cleaned of welding flux. Exposed welds should be ground smooth and flush, with weld splatter removed so that the welds blend with adjoining surfaces.

Where welding is used, connections that are not to be left as open joints should be welded. Where possible, they should be welded in the shop. Only when shipping-size limitations preclude shop welding should joints be welded in the field. Shop-paint coats should be touched up after welding, using the same paint used for the shop coat. The surfaces of exterior units that have been hot-dip galvanized after fabrication and those that will have bolted or screwed field connections or that have been factory prefinished should not be welded, cut, or abraded.

Welding of steel should be done in accordance with the American Welding Society AWS D1.1 Structural Welding Code—Steel and of aluminum with AWS D1.2 Structural Welding Code—Aluminum.

Brazing.   In the brazing process, metals are joined using a filler metal that melts at a temperature above 800 degrees Fahrenheit but at a lower temperature than the melting temperature of the metal being joined. Brazing filler metals are themselves nonferrous, but they may be used to join either ferrous or nonferrous metals. The brazing metal for aluminum is one of several alloys of aluminum. For copper alloys the brazing metal is a silver–copper brazing alloy.

Melted brazing metal distributes itself throughout a joint by capillary action, which makes brazing successful even in odd-shaped and otherwise inaccessible joints. Several methods are used in brazing, including torch brazing, furnace brazing, and dip brazing. Torch brazing is used where a joint is accessible, which makes this the only method that can be used in the field. In furnace brazing, an entire assembly is held in the furnace until the brazing metal melts. In dip brazing, the assembly is immersed in a bath of molten salt until the brazing metal melts.

Under some circumstances, brazing can produce joints that are as strong as welded ones. Because of the relatively lower temperatures involved, brazing can be used on both thin and thick metals and on castings. Brazing, not welding, is the preferred method for joining copper alloys and is also used extensively to join some types of aluminum. There are limits on the brazing of aluminum, however, because the brazing metal's melting point is close to that of the aluminum being joined. This requires accurate temperature control, which is sometimes hard to accomplish. Some alloys of aluminum cannot be brazed, because of their low melting points.

Usually brazed joints are brazed continuously. Brazing should be done on the concealed side of the joint so that the brazing metal which is generally brighter in color than the metal joined, will not be visible. Even when it is exposed, brazing metal usually requires little finishing, but the difference in color may make the appearance unacceptable. The color difference can be diminished by grinding away all excess brazing material.

Where brazing is used, connections that are not to be left as open joints should be brazed, where possible, in the shop. Only when shipping-size limitations preclude shop brazing should brazing be attempted in the field. Some metals are difficult to braze in the field, and some brazing methods cannot be used in the field. After brazing, shop-paint coats should be touched up, using the same paint used for the shop coat. The surfaces of exterior units that have been hot-dip galvanized after fabrication and that will have bolted or screwed field connections or that have been factory prefinished should not be brazed.

Soldering. In the soldering process, metals are joined by using a filler metal called solder that melts at a temperature below 500 degrees Fahrenheit. Solders are themselves nonferrous, but they may be used to join either ferrous or nonferrous metals. Solder for aluminum is usually pure zinc or zinc–aluminum. For steel, copper alloys, and stainless steel the solder is a tin–lead alloy. The 50–50 alloy usually used for steel and copper is half tin and half lead. For use with stainless steel a 60–40 solder of tin and lead is usually used. Other soldering formulations may be used, however. Some solderers will use a 70–30 solder on stainless steel, for example.

Metals must have a certain amount of roughness if solder is to bond well with them. Bright annealed stainless steel is, for example, very difficult to solder. Lead should be first scraped with a wire brush to produce a bright surface.

Some metals are difficult to solder, for other reasons. Steels with a high carbon content, for example, do not take solder well.

Because most metals that can be soldered form a thin oxide layer to which the solder will not adhere, a flux must be applied to dissolve the oxide and prevent it from re-forming. The type of flux varies with the type of metal being soldered. For steel and copper a rosin flux is usually used. For stainless steel the flux is usually an acid-chloride type, except that if the surface has been tinned (precoated with a layer of solder) the flux for stainless steel is a rosin type.

Metals to be soldered should be pretinned. The solder should then be applied slowly so that the metal becomes thoroughly heated. The melted solder will then distribute itself evenly throughout the joint, by capillary action.

Soldering is used extensively with steel, copper alloys, and stainless steel. Soldering of aluminum is possible, but it is difficult to accomplish and is best not tried in the field.

It is necessary to control solder during use and prevent or remove spills properly, and remove all traces of flux after soldering, to prevent future corrosion of the metals and permit application of a paint coating, if required. Flux will vaporize during use and condense on cold surfaces to which the flux was not applied directly. Resulting flux deposits must also be removed. The method of removing flux will differ depending on the type of flux used.

Usually soldered joints are soldered continuously.

Mechanical Joining of Components.   Joints may be made mechanically, either by attaching components together using fasteners, interlocking them along the joint, or by a process called stitching.

Fasteners.   There are many different types of fasteners used in metal fabrications. They include the general categories of bolts, rivets, stud bolts, and screws. In each group there are a myriad of types, some of which were listed earlier in this chapter in the "Materials" section under "Mechanical Fasteners." Some general procedures are applicable to them all.

First, fasteners should be concealed wherever possible. Where having exposed fasteners is unavoidable, the best type to use is countersunk, flathead Phillips-head screws or bolts.

Bolts of all types should be countersunk where they are exposed. They should be drawn up tight with their threads upset to ensure that they will not back out.

Where riveting is used, heads should be made flush where they are exposed.

Interlocking Components.   Interlocking, dovetail, and similar joints are used in heavier metals. Sheet metals are often joined using lock seams of various types. Joints that must exclude water are often soldered or filled with a sealant but many joints in interior fabrications are left unfilled.

Stitching.   Metals are sometimes joined to other metals by sewing them together with wire. This process is called stitching.

Staking.   A staked joint consists of projecting tabs on the first of two metal pieces that fit into holes or slots on the second piece to be joined. The tabs are inserted into the holes or slots, then bent or twisted to hold the pieces together.

Slip Joints.   In many fabrications, components are formed to fit (slip) over adjacent components.

Adhesive Bonding and Joining of Components.   Adhesives may be used to bond metals to other materials. For example, an aluminum sheet may be adhesive bonded to a foam board to make a wall panel.

The metal components in some fabrications may be joined together with adhesives. Copper-alloy fabrications, for example, are often assembled today using adhesives alone or combinations of adhesives and slip fittings and pins.

Provisions for Anchoring

Most metal fabrications must be anchored in place to other parts of the building construction. Their design and fabrication must therefore include provision for anchoring devices of types appropriate for use with the supporting structure. Their type, size, and spacing should be adequate to provide proper support for the use intended. Refer to Chapter 7 for additional details on anchoring metal fabrications.

The installation method to be used must be taken into account in the item's design. A fabrication that will be erected before surrounding construction is in place may, for example, have flexible anchors welded onto it. A fabrication that must be installed in an existing space will require that loose anchors be used. In addition, a fabrication that will be installed before its surrounding construction is built must be made with less dimensional tolerance than one that can be shimmed to fit.

Cleaning

Metal fabrications should be cleaned at the shop, then shipped to the construction site in a clean condition. The cleaning methods and cleaners used should not harm the metals or their finishes.

Why Metal Fabrications Fail

Some metal-fabrication failures can be traced to one or more of the following sources: concrete and steel structure failure, concrete and steel structure movement, metal framing and furring problems, wood framing and furring problems, solid substrate problems, and other building element problems. Each of these sources is discussed in Chapter 2. Although these may not be the most probable causes of utilitarian metal-fabrication failure, many of them are more serious and costly to fix than the types of problems discussed below. Consequently, the possibility that one of them might be responsible for a metal-fabrication failure should be investigated.

The causes discussed in Chapter 2 should either be ruled out or rectified if found to be at fault before utilitarian metal-fabrication repairs are attempted. It will do no good to repair existing utilitarian metal fabrications or install new ones when a failed, uncorrected, or unaccounted-for problem of the types discussed in Chapter 2 still exists, because the new installation will also fail.

General causes of failure related to metal fabrications were discussed in detail in Chapters 3 through 5 and in Chapter 7. They include material-failure problems such as corrosion and fracture (Chapter 3), finish failure problems (Chapters 4 and 5), and installation, protection, and maintenance problems (Chapter 7). When a failure has been observed, the problems discussed in those chapters should be suspected, then corrected if they are at fault.

After the problems discussed in Chapters 2 through 5 and Chapter 7 have been investigated and found not present, or if found have been repaired, the next step is to discover any additional causes for the failure and correct them.

When we try to discuss the other types of errors that can lead to failure in metal fabrications, we are limited because there are many more ways to err when designing or fabricating metal fabrications than there are types of metal fabrications—and there are hundreds of types of metal fabrications.

The number of errors possible is so enormous that here it is possible only to classify them by their types and accompany them with a few examples. A glance at the Bibliography will show that a discussion of each possible error that might lead to failure in every conceivable metal fabrication would greatly extend the length of this book. The only way to avoid design errors is to study and understand the materials to be used and how to finish, join, and anchor them. The producers of the materials and the manufacturers and fabricators of the various metal fabrications offer much help, as elucidated throughout this book. Determining the reasons for failure of a metal fabrication requires that same degree of knowledge. Fortunately, only one fabrication at a time will usually fail, so the amount of necessary research is somewhat reduced.

Some additional causes of failure include improper design and improper fabrication, which are now discussed. Refer also to the headings under "Evidence of Failure" in Chapter 7 for an indication of the types of failures that may be related to the failure causes identified here.

Improper Design

If failures are not to occur, metal fabrications must be designed properly. Improper design includes selecting the wrong materials for use in the fabrication, including the metals, fasteners, anchors, and accessories, or their supports (Fig. 6-7). One example is selecting materials that are likely to corrode for use in a corrosive atmosphere. See Chapter 3 for a discussion about corrosion failures.

Selecting metal materials that are inappropriate for a particular fabrication is an error that takes many forms. This problem includes such errors as selecting a soft metal for use where abrasion will occur, choosing a brittle metal when the components of a fabrication must be bent into curves, using sheet materials where extrusions would be a better choice, and selecting thin sheet metals for use where they will come into contact with such devices as lawn mowers, snow blowers, and the like (Figs.6-8 and 6-9).

Selecting finishes that are inappropriate for the prevailing conditions will probably result in failure of the finish. See Chapters 4 and 5 for a discussion of finishes and design errors related to them.

Designing a fabrication that is too weak for the loads to be applied to it will lead to structural failure. This error includes using metals that are too small in cross-section to support their loads. A common example is that of selecting too thin a sheet material. Another is to select structural shapes such as angles, channels, beams, and tees that are too small. Some-

Figure 6-7 The metal projection in this photo is a rain hood over vents through a metal wall (not visible here). Its condition resulted from the rotting away of wooden supports that held it up.

times a component that is ordinarily strong enough will fail because of a weakness in a particular item (Fig. 6-10).

Placing a fabrication in a location where it is likely to be damaged will often lead to failure (Figs. 6-11 and 6-12). A related problem is that of using sheet-metal fabrications too weak for contact with vehicles in locations where there will inevitably be such contact (Fig. 6-13). Sometimes a fabrication such as the fence gate shown in Figure 6-14 will be strong enough for general use but too weak for the location where it is installed.

A related error is to select fasteners or anchors that are too weak for their application, too few in number, or placed too far apart, which can lead to structural failure in the anchor or fastener. Selecting the wrong type of anchor can also lead to structural failure.

Selecting poor shapes for extrusions can produce sections that do not fit together well. There are limits on size and configuration that could prevent some designs from being made at all, but the kinds of problems that find their way into existing construction are more subtle, such as misfitting joints, and surfaces that do not align. Failure to provide sufficient

Figure 6-8 Sheet metal in locations like the one shown here often do not fare well.

and appropriate means for fastening together the components of a fabrication. Using a staked joint where a joint must move, is an example.

Requiring that incompatible fasteners be used is another. Steel fasteners will corrode if used in a copper-alloy fabrication, for example. Requiring that incompatible anchors and attachments be used is a similar problem.

Specifying that the wrong type of joining method be used can lead to failure. Requiring that aluminum components be welded or brazed in the field is one example. Requiring that a thin metal be welded to a thick one will almost always cause problems. Requiring that highly polished metals be joined by brazing will almost surely lead to a failed joint. Welding most copper alloys will produce unsightly joints. Designing open joints for use where the fabrication must exclude water will not necessarily lead to actual failure of the joint, but the fabrication will not perform as intended.

Figure 6-9 It takes only one collision to tear apart a thin metal corner such as the one shown in this photo.

Figure 6-10 This fence post has come loose because of a weakness in the collar at its top.

Figure 6-11   When the designer placed these steps where a truck wheel could run onto the lowest tread, its fate was sealed.

Figure 6-12   This photo proves that if a truck can hit a portion of a building, it will eventually do so.

Figure 6-13 This sheet-metal rain hood should have been protected from contact by trucks, or a stronger structure should have been used.

Improper Fabrication

Improperly produced metal fabrications are a sure precursor of failure. Perhaps the most frequent cause of improper fabrication is failure to follow the requirements of the design, including the shop drawings and specifications and industry standards.

Most types of fabrication errors involve joints (Fig. 6-15).

The errors include improperly making welded joints. Welding without proper preheating can lead to stress-corrosion cracking, for example. Using the wrong welding metal can result in cracking in either the weld itself or the adjacent metal.

The welding or brazing of metals, especially of thin metals from the visible side, will usually lead to unsightly joints.

Failing to properly clean a joint to be welded, brazed, or soldered will almost always result in a failed joint.

Failing to use flux before brazing or soldering a joint will probably result in failure of the brazing metal or solder to adhere to the metal, and the joint will then fail.

Another common error in producing metal fabrications is to join com-

Figure 6-14 This gate would probably have stood undamaged at the entrance to a residential yard for many years. Its use, however, was much more severe, and the damage shown became inevitable.

ponents in such a way that the fabrications are not square, plumb, and properly aligned. Joints must be flush, tight, and correctly fitted.

Using damaged materials or component parts in a fabrication, or damaging them during or after fabrication, must be avoided. Damaged components and materials should either be repaired or removed and replaced with new materials or components.

Damage to preapplied finishes such as coil coatings during fabrication can be a problem. Such damage should be touched up before the loss of protection permits the fabrication to suffer damage from rust or other corrosion. Errors in finishing are discussed in Chapters 4 and 5.

Where to Get More Information

The following publications from the Copper Development Association contain valuable information:

Figure 6-15 The top-rail joint on this section of chain link fence has failed.

- *Copper, Brass, Bronze Design Handbook: Architectural Applications.*
  A "must have" for anyone dealing with copper alloys in architectural
  (building) applications.
- *Welding Handbook.*
- "Joining Copper-Tin Alloys (Phosphor Bronzes)."
- "Joining Copper-Nickel Alloys."
- "Mechanical Properties of Soldered Copper."
- "Taking Care of the Metal in Your Building."

Harold B. Olin, John L. Schmidt, and Walter H. Lewis's book *Construction Principles, Materials & Methods* has good discussions about welding, brazing, soldering, and the mechanical fastening of steel and aluminum.

Architectural Graphic Standards, by Ramsey/Sleeper, includes much detail on metal fabrication.

There are many ASTM standards related to metal and metal fabrications. Many of those that are applicable to metal fabrications have been mentioned throughout this book. Those related to the subjects in this chapter are followed in the Bibliography with a [6].

The followng publications of the American Iron and Steel Institute contain valuable information to help the reader gain general knowledge of the subjects they cover:

- "Design Guidelines for the Selection and Use of Stainless Steel" includes information about stainless steel in general. It contains a useful table of guidelines for selecting fasteners for zinc, galvanized steel, aluminum, steel, cast iron, copper alloys, and stainless steel.
- "Welding of Stainless Steel and Other Joining Methods" is an excellent source of data about welding in general, and specifically about the welding of stainless steel.
- "Stainless Steel Fasteners: A Systematic Approach to Their Selection" and "Stainless Steel Fasteners: Suggested Source List" both offer specific guidance about stainless steel fasteners.

Timothy B. McDonald's 1989 article "Specifying Welding Details" includes some basic information about welding.

The American Welding Society (AWI) sets the standards in this country for welding. The standards listed in the Bibliography are a major source of information.

The primary source of data for someone dealing with the welding or bolting of steel is the American Institute of Steel Construction's *Manual of Steel Construction*. The other AISC entries in the Bilbiography are all included in this manual.

The Readers Digest Association's *Complete Do-it-yourself Manual* contains excellent information on many aspects of fabricating metal items. It includes data about the various ways of attaching metals, such as riveting, soldering, bolting, screwing, and using adhesives. It also contains information about forming sheet metals and repairing damaged metals. It contains a glossary of metal terms and a description of metal types.

Installation, Repair, and Cleaning of Nonstructural Metal Items

The data in this chapter are generally applicable to all nonstructural metal items (metal fabrications), as defined in this book. A partial listing of these items is included in Chapter 6. The data here may not apply, however, to structural metals or the nonstructural metal items listed as being excluded from this book under "What Nonstructural Metals Items Are and What They Do" in Chapter 1.

The support systems for the metal fabrications covered in this chapter are discussed in Chapter 2, the metals used to fabricate them in Chapter 3, the finishes for them in Chapters 4 and 5, and fabricating them and the means of joining their various components in Chapter 6.

Standards and Materials

The same codes, laws, regulations, and standards apply to the installation of metal fabrications that apply to their manufacture. Please refer to Chapter 6 for a discussion and partial listing of applicable requirements.

This section contains a discussion of the many types of anchors, inserts, fasteners, and other means of installing metal fabrications.

Anchors and Inserts

Many types of anchors and inserts are used to support metal fabrications. Some are fastened to the substrates or supporting structure with fasteners, adhesives, or welding. Others are built into a concrete supporting structure or solid substrates, such as concrete, masonry, or stone. The technique used for each metal fabrication must be the proper one for the support or substrate to which it will be fastened or into which it will be placed. It is equally important to ensure that the materials from which the anchors and inserts are made will not cause galvanic corrosion with the supported metals, fasteners, or supporting structure when it is also metal. Anchors and inserts must also be of materials that will not corrode due to their contact with the materials to which they are fastened or into which they are placed.

In general, all inserts and anchoring devices to be set in concrete or built into masonry or stone for the installation of miscellaneous metal fabrications must be furnished in time for them to be installed properly. Anchors and inserts must be of the proper size and be provided in numbers sufficient to permit properly fastening the fabrications in place.

Many brackets, anchors, and inserts used to fasten metal fabrications in place are either cast or formed metal of the same type of material and finish as that being supported. Other materials are often used, however. Cadmium or hot-dipped galvanized anchors and inserts, for example, are often used for exterior work because of their high resistance to corrosion. Stainless steel anchors may be used to anchor stainless steel or aluminum fabrications. Copper-alloy anchors are usually used to fasten copper-alloy fabrications.

Two common types of concrete inserts used to support metal fabrications are the wedge type and the threaded type. For use with ferrous metals, both types are usually galvanized castings of either malleable iron or steel, but cadmium-plated and stainless steel inserts are also used. Aluminum and copper alloys should not be used in contact with wet cementitious materials.

Where brackets and ledgers will be used to support metal fabrications, the necessary bolts, expansion bolts, expansion shields, washers, shims, and other devices should be provided. Ferrous brackets and ledgers should be hot-dip galvanized in accordance with the ASTM standards listed in Chapter 4.

Steel bar anchors are sometimes used to support metal fabrications at masonry or concrete walls. Such anchors should have turned ends and extend not less than 8 inches into masonry or 4 inches into concrete. They should be arranged to lie flat in masonry joints. Such devices should be galvanized.

Fasteners

Many types of fasteners are used to anchor metal fabrications in place, depending on the material to which the fabrication is to be attached. Some of the fasteners used include machine bolts, carriage bolts, stove bolts, toggle bolts, molly bolts, hanger bolts, lag bolts, expansion bolts, and nails, screws, screw hooks, clamps, sockets, and sleeves. Each bolt should have the proper matching nuts and washers. Some anchors also require expansion shields or various types of wall plugs. The necessary items are usually supplied with their fabrications.

The fasteners used to anchor metal fabrications should be concealed when practicable. When fasteners must be exposed, they should be of the same material, color, and finish as the metal being fastened. For instance, aluminum or stainless steel fasteners should be used in aluminum, and fasteners used in color-anodized aluminum should match the aluminum finish. Painted fasteners should not be used in color-anodized aluminum, because they will almost never match the aluminum's color and will be immediately and objectionably apparent.

Stainless steel fasteners are often used to fasten stainless steel, chromium-plated, and aluminum items in place. Neither carbon steel nor aluminum should be used in conjunction with copper alloys, since they will corrode readily in contact with most copper alloys. If water is present, even stainless steel may corrode when in contact with copper alloys. Therefore, copper-alloy fabrications should not be fastened to steel or cast-iron inserts or to any other materials using steel or aluminum fasteners.

If the fabrication to be installed is a ferrous metal, then zinc-coated or cadmium-plated steel or stainless steel fasteners should be used in exterior locations and where they are to be inserted into an exterior wall.

In every case, fasteners should be of the type, grade, and class appropriate for the items being fastened. The following list contains a few of the many fasteners and related devices used to fasten metal fabrications in place. Following each item is the standard with which that item should comply.

- Bolts and nuts: ASTM Standard A 307, Grade A, or Federal Specification FF-B-575C.
- Lag bolts, square-head type: Federal Specification FF-B-561.
- Machine screws, cadmium-plated steel: Federal Specification FF-S-92.
- Wood screws, flat-head carbon steel: Federal Specification FF-S-111.
- Plain washers, round, carbon steel: Federal Specification FF-W-92.

- Expansion shields: Federal Specification FF-S-325.
- Toggle bolts, tumble-wing type: Federal Specification FF-B-588 of the type, class, and style required.
- Lock washers, helical spring-type carbon steel: Federal Specification FF-W-84.

Where bolting is used, the proper-sized bolts should be installed. Their heads should be concealed where possible and countersunk where exposed. Bolts should be drawn up tight and their threads upset to ensure that they will not back out.

Screws should be concealed, but if this is impossible, the screw heads should be countersunk. The type of screw usually preferred for use as exposed fasteners is Phillips flat-head screws.

Integral Anchors

Some metal fabrications will have integral anchors formed into or welded onto the fabrication. Such fabrications must therefore be installed either before the surrounding construction is erected or while it is being built. Integral anchors are formed from the same material as the metal fabrication, while the item is being fabricated (see Chapter 6).

Grout

Nonshrinking grout is often used to seat and level metal fabrications. The two most-used types, metallic and nonmetallic, either do not shrink, shrink so little as not to affect their use, or actually expand while curing.

Metallic nonshrink grout is a premixed, factory-packaged material containing a ferrous aggregate. It should comply with U.S. Department of the Army CRD-C621 "Specification for Nonshrink Grout."

Nonmetallic nonshrink grout is a premixed, factory-packaged nonstaining, noncorrosive, nongaseous grout also complying with U.S. Department of the Army Specification CRD-C621.

The grout used should be specifically recommended by the manufacturer for interior or exterior use, in the conditions that exist where the grout is to be used.

Anchoring Cement

A typical use for anchoring cement is to set a steel pipe railing into a sleeve. Until relatively recently, most such installations were made with molten lead (Fig. 7-1). This method is still used today, especially in exterior locations, but not as extensively as previously, because of the development

Figure 7-1   A tubular railing standard set with lead. This installation has begun to fail and should be repaired soon.

of effective anchoring cements. Different types of these cements are used in exterior and interior applications. Both types are nonstaining.

Interior types of anchoring cements come from the factory ready for mixing with water at the construction site. They work as intended because they do not shrink, and may even expand, as they cure.

The exterior types are similar to interior ones but are, in addition, resistant to erosion. These materials should be those specifically recommended by their manufacturer for use on exteriors (Fig. 7-2).

Miscellaneous Materials

Chapter 6 included descriptions of some materials and items that are also used during field assembly of metal fabrications that could not be moved into the installation location in one piece. They include bituminous paint; welding and brazing filler materials, solder, and associated materials; mechanical fasteners; sleeves; insulation; and metal-to-metal sealants.

Compression Seals.   Where metal fabrications must prevent the passage of air, it is often necessary to provide some sort of compression seal between

Figure 7-2   This photo shows the result of selecting the wrong type of anchoring cement, which is a design error.

the various parts of a fabrication or between the fabrication and adjacent materials. One such commonly used seal is gaskets made from closed-cell polyvinyl chloride (PVC) foam strips.

Concrete for Stair Treads and Landing Fill.   Concrete for stairs and landings is usually normal weight, ready-mix concrete with a minimum 28-day compressive strength of 2,500 psi, although higher-strength concrete is occasionally used.

Sealants.   The type of sealants best used between metal fabrications and adjacent materials depends on the type of fabrication and its adjacent material. The sealants generally used are elastomeric ones conforming with the requirements of ASTM C 920. The actual type used depends on the location. Acrylic sealants are also used. Sealants and their uses are beyond the scope of this book, but some insight can be gained into their selection and use by studying the references listed in "Where to Get More Information" at the end of this chapter.

Finishes

A discussion of the types of finishes normally used on the metal portions of metal fabrications is included in Chapter 6. Chapters 4 and 5 contain detailed information about those finishes.

Many metal fabrications, especially those made from iron and steel, are first given a shop coat of primer paint, then field painted, as discussed in Chapter 5.

When a second coat of paint has not been applied in the shop on portions of metal fabrications that will be inaccessible after assembly or erection, it should be applied to such areas in the field before the fabrication is installed. Shop coats of paint that have been applied inadvertently on portions of metal fabrications that are to be embedded in concrete, stone, or masonry, as well as from surfaces that will be field welded and from galvanized surfaces, should be removed.

Fabrication

The general requirements for fabricating nonstructural metal items were discussed in Chapter 6. Often, metal fabrications cannot be completely assembled in the shop and must be shipped to the construction site disassembled. When field assembly is necessary, the principles discussed in Chapter 6 generally apply. There are, however, a few additional requirements that must be observed during field assembly. It is, for example, often much easier to assemble metal fabrications under shop conditions. Aluminum and some other metals that can be welded under the controlled conditions possible in the shop are very difficult to weld in the field. This section discusses further requirements for field fabrication.

Field joints should be kept to the minimum number possible. Care should be taken when making them that their rough edges are concealed or further finished to eliminate roughness. The final dimensions after assembly should be those shown on the approved shop drawings. Proven methods of assembly and support should be used.

As is true in the shop, field-welded corners and seams should be welded continuously. Welding should comply with AWS recommendations. Welds should be made behind finished surfaces, without distortion or discoloration of the exposed side. The exposed welded joints should be cleaned of welding flux. Exposed welds should be ground smooth and flush, so that they blend with adjoining surfaces.

Joints should be pulled up to a tight hairline fit. Exposed connections should be made flush and smooth, using concealed fasteners wherever possible. If using exposed fasteners is unavoidable, the best type to use in most cases is countersunk flat-head Phillips-head screws or bolts. Metal panels of a more utilitarian than decorative nature (Fig. 7-3) are often joined with rivets or other fasteners, however.

The type of anchorage should be the proper one for use with the supporting structure. The type, size, and spacing of anchors should be adequate to provide proper support for the intended use.

Fabrications should be installed so as not to defeat their designed-in allowance for expansion and contraction (See Chapter 6).

Matching trim should be installed where necessary to complete the installation.

Installing New Fabrications in New Buildings

The information here applies as well to installing new fabrications in an addition to an existing building, since both the fabrication and its surrounding and supporting construction are new.

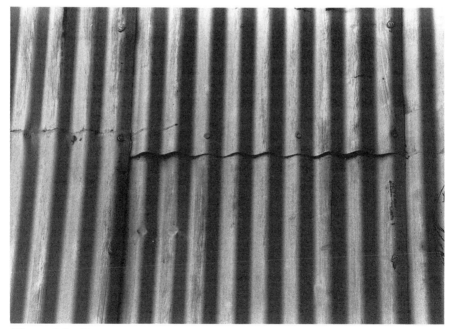

Figure 7-3   Rivets have been used to join these corrugated metal panels in the field.

Preparation for Installation

For proper installation of utilitarian metal fabrications, certain preparations must be made.

Attachments and Anchors.   Sleeves, anchorages such as concrete inserts and anchor bolts, and miscellaneous items with integral anchors that are to be embedded in concrete or masonry must be delivered to the project site at the proper time. Devices to be cast or set in must be delivered before the materials into which they are to be built or cast have been placed.

Actual Conditions at the Site.   The location where metal fabrications are to be installed should be examined carefully by the installer. This examination should verify that necessary adjacent and supporting construction is in place and that the areas and conditions where the metal fabrication will be installed are in a proper, satisfactory condition to receive the new metal fabrication. Preparatory work should not be started and the metal fabrication should not be installed until unsatisfactory conditions have been corrected.

The installation of metal fabrications must be coordinated with the work of the mechanical, electrical, and other trades to avoid interference with or damage to concealed pipes, vents, ducts, and conduit and to ensure that proper support is available for the new metal fabrication. Reinforcement must be provided in stud walls where handrail brackets occur, for example, and toilet partitions must be reinforced to support handrails for the handicapped. The sizes and locations of openings in other construction where metal fabrications will be placed must be coordinated with the fabrications. For instance, toilet partitions must have proper openings to receive recessed toilet-tissue holders.

The fabrications must be delivered and installed at the proper time, because delivering a metal fabrication too soon may block the installation of other construction. Delivering too late may delay the building of surrounding construction or prevent installation of a piece of equipment or other item that must be supported by the metal fabrication.

Installation

Damaged metal fabrications should not be installed, because it is generally much easier to repair a damaged fabrication before it is installed. Often, damage will not be repairable after installation, and the fabrication must then be removed and a new unit provided. Damaged metal items that have been installed should be removed and repaired, or new, undamaged units should be installed. Installing already damaged metal fabrications is one of the largest causes of later problems with metal fabrications.

Requirements Applicable to Installing All Metal Fabrications. Metal fabrications should be installed in accordance with the manufacturer's latest published instructions and the approved shop drawings, and be securely anchored in place. There will be slight differences in the installation methods available depending on whether the fabricated unit is to be installed before the space is built, as the space is being erected, or if it is to go in an already prepared space. The first two situations are usually possible, and in some cases customary, in all-new construction projects. Steel stairs, for example, are sometimes the first element visible above ground at an apartment building site.

There are three major advantages in making use of the first two installation methods. First, they make it possible to build anchoring and supporting devices, such as anchors and inserts, into new concrete, masonry, and stone as it is erected. Second, they permit shop fabrication and delivery of the metal unit to the construction site in large sections, often in one piece. Third, they make extreme accuracy in fabricating the metal units somewhat less important, because the surrounding construction can be slightly adjusted during construction. This can prove a problem, however, if the fabricator takes advantage of the leeway to give too little attention to accuracy. The result can be aesthetically less than pleasing.

Even in new buildings, metal fabrications are often installed after the surrounding construction has been erected. One advantage of this method is that fitting the fabrication in place can be done more accurately when measurements can be made of the space to receive the fabrication, and the fabrication can be made accurately to fit the space. Another advantage is that placing the fabrication in its proper location accurately is easier.

The types of anchors used may vary, of course, depending on when in the construction process a fabrication is installed. A metal fabrication to be anchored to concrete or masonry that has not yet been constructed can be anchored using threaded fasteners placed into inserts or to fasteners set into masonry or stone joints. If the masonry, stone, or concrete are already there, then toggle bolts, through bolts, threaded fasteners into expansion shields in drilled holes, and similar fasteners and anchors will be necessary.

In order to install metal fabrications it is often necessary to perform cutting, drilling, and fitting of either the fabrication or the adjacent construction. Such work should be done so that the fabrications can be set accurately, in the proper location, alignment, and elevation. Fabrications should have their surfaces and edges plumb, level, true, and free of rack, when measured from established lines and levels.

In most fabrications, exposed connections should fit together accurately to form tight hairline joints. The best field connections of pipes and tubes are made using internal sleeves. Connections that are not to be left as open joints should be welded, in the shop where possible. Where shipping-size limitations preclude shop welding the joints should be welded in the field. In either case exposed welds should be ground smooth. Paint coats should be touched up after welding, using the same paint that was used for the shop coat. The surfaces of exterior units that have been hot-dip galvanized after fabrication and that will have bolted, riveted, or screwed field connections or have been factory prefinished should not be welded, cut, or abraded. Refer to Chapter 6 for additional suggestions about welding.

When the joints between metal fabrications and adjacent surfaces are not tight, they should be trimmed or filled with sealant. Before sealant is applied, strippable coatings and other applied protective materials must be removed.

Metals, especially aluminum and copper-alloy surfaces, should be protected as recommended in Chapter 3 from corrosion where they are in contact with dissimilar metals, concrete, masonry, or pressure-treated wood.

Specific Requirements for Installing Accessories and Specialties. Some special requirements apply to metal fabrications that are manufactured units. Although not all such devices are called accessories or specialties in the building industry, the basic principles discussed here are applicable to the others as well. The types of accessories and specialties we are talking about here include toilet accessories, identifying devices, expansion-control devices, chalkboards, tackboards, and many others. What we have to say will apply as well to hardware, ornamental handrail brackets, pass-through slots, postage accessories, and many other devices.

Most large accessory and specialty items, and all small ones, are customarily delivered to a building site carefully packaged, complete with the necessary mounting devices and hardware, to prevent damage and loss. Owners often require that a contractor provide a locked, protected place to store small accessories and specialties until they are installed, to protect them from damage or theft.

Accessories and specialties should be complete and ready for use, with all the necessary operating parts, hardware, and accessories.

The necessary grounds, back blocking, inserts, bolts, screws, and other accessory items should be provided to permit suitable anchorage to supports.

Fastenings should be solid, substantial, and strong enough to meet the requirements of normal use. Concealed fasteners should be used when possible. Units should be attached securely to walls and partitions in their proper locations. Exposed mounting devices and fasteners should be used only where concealing them is not possible. Their finish and material should match the accessory or specialty. Most fasteners should be theft resistant, and in public places all should be. Recessed accessories and specialties in stud walls should be fastened to the studs or to horizontal framing between studs. Surface-mounted units should be fastened through the finish to the underlying substrate or framing, never to the supported finish alone.

Cleaning Installed Metal Fabrications

Metal surfaces should be kept clean during construction and cleaned at the completion of construction to remove mortar, paint, dirt, grease, dust, efflorescence, salts, and stains. The cleaning materials used should not harm the metal, metal finishes, or adjacent surfaces. Abrasives and caustic or acid cleaning agents should not be used. Cleaning methods and materials should follow strictly the recommendations of the manufacturers of the metal and its finish. The normally recommended cleaning for aluminum, for example, permits cleaning the metal initially with soft rags or brushes, using only clear water and mild soap. Only if the initial cleaning does not remove soiling may the aluminum then be washed with mineral spirits, washed with mild soap and clean water, rinsed with water, and allowed to dry. The producers of other metals and finishes will offer similarly specific recommendations for cleaning their products.

Immediately after a fabrication has been erected, its field welds, bolted connections, and areas of abraded shop paint should be cleaned, and exposed areas painted with the same material used for the shop painting. Touch-up paint should be applied by brush or spray so as to provide a minimum dry-film thickness of 2.0 mils.

Field welds, bolted connections, and abraded areas in galvanized surfaces should be cleaned and two coats of galvanizing repair paint applied.

Metal fabrications that are to be field painted should be further prepared and painted in accordance with the general recommendations of the paint manufacturer (see Chapter 5).

Upon completion of the construction work, temporary masking, strippable coatings, and other protection should be removed from metal fabrications. If completion is delayed, it may be necessary to remove such protection (especially strippable coatings) earlier, since some of them may be impossible to remove completely after they have been exposed too long to heat or sunlight.

Why the Anchoring of Metal Fabrications Fails

Some anchorage failures can be traced to one or more of the following sources: concrete or steel structure failure, concrete or steel structure movement, metal framing and furring problems, wood framing and furring problems, solid substrate problems, or other building element problems. Each of these sources was discussed in Chapter 2. Although many of these are perhaps not the most probable causes of metal fabrication failure, they are more serious and costly to fix than the types of problems discussed here. Consequently, the possibility that one of them may be responsible for a metal fabrication's failure should be investigated.

The causes of failure discussed in Chapter 2 should either be ruled out or rectified if found to be at fault before repairs to a metal fabrication are attempted. It will do no good to repair existing metal fabrications or install new ones when a failed, uncorrected, or unaccounted-for problem of the types discussed in Chapter 2 exists, because the new installation will also fail.

General causes of failure related to metal fabrications were discussed in detail in Chapters 3 through 6. They include material failure problems such as corrosion and fracture (see Chapter 3), finish failure problems (Chapters 4 and 5), and improper design or fabrication (Chapter 6). When a failure has been observed, the problems discussed in those chapters should be suspected, and corrected if at fault.

After the problems discussed in Chapters 2 through 6 have been checked for but found not to be present or have been repaired if found, the next step is to discover any additional causes for the failure and correct them.

Two of these additional causes are bad installation workmanship and failing to protect the installation. The following discussion covers these additional causes. Refer to "Evidence of Failure" later in this chapter for an indication of the types of failures that may be related to the causes of failure identified throughout this book.

Bad Installation Workmanship

Correct installation is essential if metal fabrication failures are to be prevented.

Probably the number-one cause of failure from installation errors in metal fabrications is a failure to follow the design and the recommendations of the fabricator (manufacturer) and recognized authorities, such as the ASTM, AAMA, or NAAMM. One major type of error that can occur when the design and standards are ignored is the use of incompatible metals or of a metal anchor that will corrode when in contact with wet concrete or masonry mortar.

Failing to follow the design can mean using anchors, fasteners, and other materials that will rust where high humidity will occur or where they are set in or are in direct contact with masonry or concrete.

It can also lead to using anchors, fasteners, clips, and other devices of the wrong type for an installation, or that are too small or too weak to perform the necessary function, or that are not in the proper number or location.

Failures can also occur if the area where a metal fabrication is to be installed is not properly prepared or ready to receive the fabrication. Installing metal fabrications before a building has been completely closed and wet work such as concrete, masonry, and plaster have dried out sufficiently can lead to corrosion in the metal fabrications. Placing a load on newly set anchors by attaching the metal fabrications before the concrete or masonry mortar surrounding them has set, can result in loose anchors, which may come completely out of the wall. A related error is installing metal fabrications so that building loads that the fabrication is not designed to support are transmitted into the fabrication.

An allied error is installation of a damaged metal fabrication. Although the damage may or may not be the fault of the installer, putting a damaged fabrication in place certainly is.

An installer must be careful to install metal fabrications so that their various components finish in the proper plane within acceptable tolerances and the units are in the proper location—plumb, level, and in alignment. Improperly aligning a surface may not only cause the fabrication to present an unpleasing appearance but can force loads to be applied eccentrically, which may cause the fabrication to become damaged or even collapse.

One installation error is failing to seal around the perimeter of a metal fabrication, which will leave a poor-appearing installation.

Installing a metal fabrication using an adhesive in a space that does not meet the temperature and humidity requirements of the fabricator or of a standard that is applicable is an error. Some adhesives are adversely affected by temperatures or humidities that are even slightly high. A similar problem will occur if an adhesive is applied to a hot or wet substrate or the metal fabrication is hot or wet when the adhesive is applied.

Failure to Protect the Installation

Metal fabrication components must be protected before, during, and after installation. Related errors include failing to protect metal fabrications from staining, denting (Fig. 7-4), scratching, or other damage from other con-

Figure 7-4 This metal siding has reached the condition in this photo by repeated contact with trucks. Even after the damage was initially observed, no steps were taken to prevent its reoccurence.

struction materials or procedures. Permitting abuse before or during installation or after the metal fabrication has been installed is a further error.

Some metal fabrications will fail even when they are seemingly well protected, as seen in Figure 7-5.

Evidence of Failure

In the following discussion metal fabrication failures are divided into such failure types as "Bent, Twisted, or Deformed Metal Fabrications." Each failure type is listed under its own heading. The list following each failure type indicates the location in this book where the causes of that kind of failure may be found.

A discussion of the errors that can lead to each type of failure listed can be found in the chapter whose number follows the listing, under a heading with exactly the same name. To make them easier to locate, each listed failure cause is included in the Contents.

Thus, to discover a possible cause for a damaged finish on a metal

Figure 7-5 The railing shown in this photo was regularly painted, but it corroded where it could not be painted, below the lead used to anchor it into a steel sleeve set in the concrete. The stains on the concrete are rust.

fabrication, look under "Bent, Twisted, or Deformed Metal Fabrications." One possible cause listed there is "Concrete and Steel Structure Failure: Chapter 2." Open Chapter 2 to "Concrete and Steel Structure Failure." Examine the causes of failure discussed there, to discover if any of them may be applicable to the type of damage being observed. If none seems to correlate with the damage, look under the next possible cause and so on, until the actual cause is identified. Of course, there may be more than one cause for a particular failure.

Rust, Corrosion, or Damaged Finish. A metal fabrication can become rusted or corroded in various ways, or its finish can be damaged in many ways and at many levels, from minor scratches to the complete delamination of an applied finish (Fig. 7-6). Many different types of damaged-finish problems and their causes were discussed in Chapters 4 and 5. When a damaged

Figure 7-6 The metal in the fence hinge shown in this photo is corroding. If left unrepaired, it will eventually corrode completely away.

finish is observed on a metal fabrication, search for a cause in one of those chapters or in the other chapters to which the references there may lead.

Bent, Twisted, or Deformed Metal Fabrications.   Metal fabrications may be damaged before, during, or after installation (Figs. 7-7 and 7-8). The reasons for such damage are discussed under the following headings:

- Concrete and Steel Structure Failure: Chapter 2.
- Concrete and Steel Structure Movement: Chapter 2.
- Metal Framing and Furring Problems: Chapter 2.
- Wood Framing and Furring Problems: Chapter 2.
- Solid Substrate Problems: Chapter 2.
- Errors That Lead to Fracture: Chapter 3.

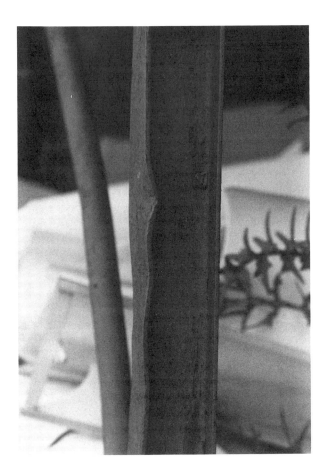

Figure 7-7 It might be possible to straighten this damaged channel, but since it is in a utilitarian area repairing it is probably not worth the effort.

- Improper Design: Chapter 6.
- Improper Fabrication: Chapter 6.
- Bad Installation Workmanship: Chapter 7.
- Failure to Protect the Installation: Chapter 7.

Light-colored Bands, Cracks, and Breaks.   Often a change of color is the first sign that fracture of a metal is imminent. The next phase is often a fine hairline splitting, frequently with rays emanating from the split. The last phase of fracture is, of course, separation of the metal. Fracture may be caused by one of the situations discussed below:

- Concrete and Steel Structure Failure: Chapter 2.
- Concrete and Steel Structure Movement: Chapter 2.

Figure 7-8 Like the damage shown in Figure 7-7, the damaged wall panel in this photo is in an area where its appearance is not too important. Nevertheless, this damage must be repaired, because it lets water enter the building.

- Metal Framing and Furring Problems: Chapter 2.
- Wood Framing and Furring Problems: Chapter 2.
- Solid Substrate Problems: Chapter 2.
- Errors That Lead to Corrosion: Chapter 3.
- Errors That Lead to Fracture: Chapter 3.
- Improper Design: Chapter 6.
- Improper Fabrication: Chapter 6.
- Bad Installation Workmanship: Chapter 7.
- Failure to Protect the Installation: Chapter 7.

Loose, Sagging, Fallen, or Out-of-Line Metal Fabrications. Metal fabrications may become loose and sag or fall off a wall if the anchors or fasteners attaching it to the wall fail. This type of failure may occur because the structure or other support system fails or because the anchors or fasteners fail. Sometimes such a failure will occur when the fabrication itself is not strong enough to support its own weight. There are many possible reasons for such a failure, which may be found under the following headings:

- Concrete and Steel Structure Failure: Chapter 2.
- Concrete and Steel Structure Movement: Chapter 2.
- Metal Framing and Furring Problems: Chapter 2.
- Wood Framing and Furring Problems: Chapter 2.
- Solid Substrate Problems: Chapter 2.
- Other Building Element Problems: Chapter 2.
- Errors That Lead to Corrosion: Chapter 3.
- Errors That Lead to Fracture: Chapter 3.
- Improper Design: Chapter 6.
- Improper Fabrication: Chapter 6.
- Bad Installation Workmanship: Chapter 7.
- Failure to Protect the Installation: Chapter 7.

Repairing Existing Metal Fabrications

Existing metal fabrications that have been damaged or are in an unsightly condition but still usable may often be repaired and restored. Sometimes it is necessary to remove the fabrication to effect repairs, but many can be made without removing the unit.

When working with existing metal fabrications it is best, when possible, to use materials and methods that match those in the original installation. When a part of a fabrication has been damaged, it may be necessary to remove and discard that part and provide a new replacement one. When fasteners or attachments have failed, it will be necessary to provide new attachments. Metal fabrications that are damaged too severely to repair, of course, must be discarded and new units provided.

The number of possible conditions for damages is far too numerous even to list, much less describe in detail in any single book. Sometimes, an apparent failure in a metal fabrication is actually failure of an adjacent material (Fig. 7-9). Each condition must therefore be examined in light of the data included not only in this book but also in the referenced sources for additional information about new fabrications. The condition of the damaged fabrication should then be brought as close as possible to the condition usually required for that type of item when new.

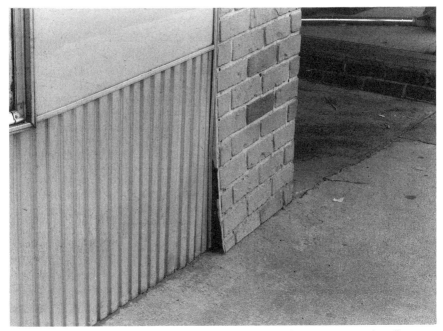

Figure 7-9   At first glance the metal panel in this photo seems to be distorted. The actual distortion is, however, in the fake brick adjacent to the corrugated metal.

Support failures, finish failures, fractures, fastener and attachment failures, and other types of failures are examined generically in Chapters 2, 3, 4, 5, and 6 and this chapter. References to sources of additional information are included both within and at the end of each chapter. When a certain type of damage is observed, it is essential to muster all available resources: the specific data in all of the chapters, including this one, the data in the sources listed for additional information, and the expertise of someone knowledgeable about the type of problem that exists.

In addition to the requirements for repairs discussed earlier, where existing metal fabrications must be repaired in place or removed, repaired, and reinstalled, there are additional requirements. At least some selective demolition, dismantling, cutting, patching, and removal of existing construction is usually necessary. The comments related to such matters later in this chapter under "Installing New Fabrications in Existing Spaces" apply also, when relevant, to an existing fabrication that needs to be repaired.

General Requirements

The following discussion contains suggestions for repairing existing metal fabrications. Because these suggestions are meant to apply in many situations, they might not fit a specific case. In addition, there are many possible conditions that are not specifically covered here. When a condition arises in the field that is not addressed here, advice should be sought from the additional sources of data mentioned throughout this book. Often, consultation with the fabricator or manufacturer of the materials being repaired will help, but sometimes it is necessary to obtain professional help (see Chapter 1). Under no circumstances should the specific recommendations in this book be followed without careful investigation and the judicious application of professional expertise.

Before an attempt is made to repair an existing metal fabrication, the existing fabrications fabricator's (manufacturer's) and finisher's recommendations for making the repairs should be obtained. It is necessary to be sure that the recommended precautions against materials and methods that may be detrimental to the fabrication or its finish are followed.

Unless existing metal fabrications must be removed to facilitate repairs to them or to other portions of a building, or they are scheduled to be replaced with new materials, only sufficient materials should be removed as is necessary to effect the cleaning and repairs. The more materials that are removed, the greater the possibility of additional damage occurring during the handling and storage of them. The materials to be removed should include those that are loose, sagging, or damaged beyond in-place repair. Other components that are sound, adequate, and suitable for reuse may either be left in place or else removed, cleaned, and stored for reuse.

After metal fabrications or their components have been removed, concealed damage to such supporting elements as framing, furring, or substrates may become apparent. Such damage should be repaired before repaired or new metal fabrication components are installed. Refer to Chapter 2 for a discussion of repairing wood and metal framing and furring. The repair of concrete and steel framing systems, solid substrates, and other building-element problems is beyond the scope of this book.

Existing metal fabrication components and supports that are to be removed should be removed carefully, and adjacent surfaces should be protected so as not to damage them. The various materials and components should be removed in small sections. Where necessary to avoid the collapse of construction that is not to be removed, temporary supports should be installed. Removed materials that are to be reinstalled should be handled carefully, stored safely, and protected from damage. Debris should be removed promptly so that it will not be responsible for damaging materials that are to remain in place.

Unless the decision is made to discard damaged metal fabrication and support-system components, ones that can be satisfactorily repaired should be, whether they have been removed or left in place. Failing to repair known damage may lead to further failure later on. The methods recommended by the fabricator (manufacturer) of the materials should be followed carefully when making repairs, even ones as simple as touch-up painting. Metal fabrication and support-system components that cannot be satisfactorily repaired should be discarded and new matching materials installed.

Areas where repairs are to be made should be carefully inspected to verify that existing components to be removed have been and that the substrates and structure are as expected and not damaged. Sometimes, substrate or structure materials, systems, or conditions are encountered that differ considerably from those expected, or unexpected damage is discovered. Damage previously known as well as that found later should be repaired before metal fabrication components are reinstalled. Reinstallation should not proceed until other work to be concealed or made inaccessible by it has been completed and unsatisfactory conditions have been corrected.

New materials should be provided where patching of an existing metal fabrication is required and the necessary materials do not exist or existing materials are not acceptable for reuse, or where the existing materials are in insufficient quantity to complete the work. The new materials should match the originally installed and removed materials in type, size, material, edge condition, thickness, finish, and other characteristics. The new installation should match exactly that used in the original fabrication. Substrates should be repaired and cleaned as necessary to obtain a satisfactory installation.

Patched fabrications should be complete, with no voids or openings. The patches should be made to be as inconspicuous as possible.

Repairing Specific Types of Damage

In addition to the general requirements for repairs that have already been discussed, the following specific recommendations apply to specific problems.

Rust, Other Corrosion, or Damaged Finishes.   The best method of repair will depend on the type of material and the extent of the damage to it. Scratched and abraded painted surfaces can be repainted using any paint type recommended for use on metal surfaces. Preparation should be as recommended by the paint's manufacturer. Refer to Chapter 5 for a discussion of paints, including materials, surface preparation, and the use of materials. The paint used must be compatible with the existing finish. Should there be any doubt about the compatibility of materials, it is best to contact the paint manufacturer for advice. Except for minor touch-ups of scratches, it is usually better to refinish an entire member rather than just the damaged area, so that the repair will be less apparent.

Minor scratches, nicks, and chips in painted metals may be concealed using chalk, pastels, typists' correction fluid, shoe polish, or a small amount of paint.

Clear coatings on natural aluminum finishes can usually be repaired using the same clear coating as originally applied, but such repairs may be visible. These clear coatings are usually, but not always, of lacquer. See Chapter 5 for further discussion.

The repair of anodized aluminum, plastic coatings, fluoropolymer finishes, and baking-finish enamels should be carried out in accordance with the manufacturer's and finisher's recommendations. Similar coatings may not respond alike to the same refinishing method. Techniques that work well with one finish may be harmful to a similar one. Refer to Chapters 4 and 5 for additional information.

Damaged finishes on metal members that are not exposed may be repaired using the same type of material as in the original finish. Painted finishes may be repainted. Galvanized surfaces can be recoated, using galvanizing repair paint. When a concealed metal has begun to fail, as happened to the tubular standard in Figure 7-10, the only solution is to remove the fabrication and repair and reinstall it. In situations such as that shown in Figure 7-10, the standard will be difficult to remove. The tendency, therefore, is to cut the standard and drill out the remaining piece of it and its anchoring material. Then a short piece of standard is anchored in the old sleeve and a new, shorter standard welded to the short piece. This is a terrible solution. It will look ugly and it will probably fail again, because the welding will burn away the protective coating from the concealed portion of the new standard. The correct solution is to carefully remove the old anchoring material, repair the standard if possible, and reinstall the entire fabrication, using methods that will prevent the original problem from recurring. The real problem shown in Figure 7-10 is that the anchoring material was not brought higher than the sleeve and sloped to drain water away from the standard or, alternatively, caulked to achieve the same results. If the old standard cannot be satisfactorily repaired and coated for protection, it should be discarded and a new one provided.

Bent, Twisted, or Deformed Metal Fabrications.   If the damage is not too severe, it may be possible to remove dents and straighten twisted or deformed elements in metal fabrications (Fig. 7-11). A slight dent in thin metal can often simply be pushed back into place if the area is accessible from the opposite side. When the back side is accessible, but the dent resists hand pressure, it can sometimes be beat back into shape with a

Figure 7-10 The railing standard in this photo has corroded below the level of the sleeve into which it is set. The corrosion has reached a stage where it has extended above the top of the sleeve and destroyed the paint finish there.

leather or wooden mallet. A sandbag held against the finished side of the area being bent back into shape will usually help make the rebent metal smoother. Sometimes dents in thin metal can be pulled back into place using simple suction tools, such as a plumber's helper.

Small dents in heavier metal can often be filled with epoxy or another filler, then ground down flush with the surrounding surface. Even large dents in thick metals can sometimes be filled with epoxy. It may be possible to pull out a large dent in thick metal by inserting screws into it and pulling on the screws.

Holes in metal can be filled using an epoxy mixture applied against a screen wire held fast to the back side of the metal. Most metal repairs, however, especially those involving bent or twisted metal, are not do-it-yourself projects. Expert knowledge and equipment are needed. Often the

Figure 7-11 The bent wrought iron spike shown can probably be straightened by an experienced metalworker. Proper techniques must be followed. Heating the metal, for example, will probably be necessary. As simple as this looks, it is not a job for an amateur.

damage cannot be repaired until the metal fabrication has been removed (Fig. 7-12). Often it is necessary to send the item to a shop to make the repairs. Rebending a metal may set up stresses in it that can be relieved only by heat treatment of some sort, which is usually impracticable to do in the field. The necessity of removing a fabrication to a shop for expensive repairs sometimes makes the process so expensive that it is cheaper to discard the damaged item and provide a new one.

Sometimes, straightening a badly damaged metal fabrication component is impracticable, even when cost is not an object. In this case, removing the damaged component and providing a new one may be the only alternative.

Loose, Sagging, Fallen, or Out-of-Line Metal Fabrications. A metal fabrication may become loose, sag, fall off the wall, or appear to be out of line because it was poorly designed or fabricated. Sometimes the metal used will simply be too thin for the item to support itself. It is difficult to fix a badly designed metal fabrication, although a too-thin section can

Figure 7-12 Repairing the damaged sheet metal shown in this photo may not be possible under any circumstances. It will almost certainly not be possible to repair this without removing the damaged metal so that it is accessible from the concealed side.

sometimes be supported by a secondary member. In other cases it may be necessary to completely shore up the fabrication, which usually means shipping it to a shop.

Most of the time, when a metal fabrication is loose, sagging, out of line, or falling away from its support the fault lies in the anchoring system. Such problems may be relativley easy to solve. New anchors in the right locations will sometimes do wonders. The appearance of such a finished piece, however, will not always be good. Since the existing anchors were in the wrong places or were inadequate in size or number, it will usually be necessary to install the new anchors where they will be visible.

Missing Components.   It is usually possible to find or make replacements for missing components (Figs. 7-13 and 7-14). Often a new component can be installed quite easily in the field (Fig. 7-15).

Repairing Materials Adjacent to Metal Fabrications.   When damage to a metal fabrication has affected adjacent construction, repair of that construction is also required.

Figure 7-13 The missing spike on this fence will probably not be hard to replace, since a duplicate can be made fairly easily.

Figure 7-14 The missing decoration on this fence will be more difficult to find than the missing piece in Figure 7-13. This one may require that a special mold be made, unless the original mold is still available or a matching stock element can be found. The latter, while unlikely, is still often possible.

7-15 Making a sheet-metal panel to replace the missing one shown in this photo is probably the hardest part of the job. Installation should literally be a snap.

Cleaning of Existing Metal Fabrications

The same principles apply to cleaning existing metal fabrications that apply to cleaning new fabrications, which was discussed under "Cleaning Installed Metal Fabrications" earlier in this chapter. Because the methods suggested are meant to apply to many situations, they might not apply to a specific case. In addition, there are many possible cases not specifically covered here. When a condition arises in the field that is not addressed here, seek advice from the additional sources of data mentioned in this book. Often, consultation with the producer or finisher of the materials being cleaned will help. Sometimes it is necessary to obtain professional help (see Chapter 1). Under no circumstances should the specific recommendations in this book be followed without careful investigation and the judicious application of professional expertise.

Before an attempt is made to clean existing metal fabrications, the existing material's producer's and finisher's recommendations for cleaning should be referenced. Be sure that the manufacturer's recommended precautions against using materials and methods that may be detrimental to finishes are followed. Because the suggestions in this book are general in nature they should not take precedence over instructions from the material's producer or finisher.

On most painted metals, soiled areas may be washed using a mild soap and a damp cloth or sponge. Detergents and soaps containing a detergent should not be used. Some types of finishes may withstand more scrubbing than others without damage, but each should be tested in a small, inconspicuous area first. It is not always prudent to accept blindly the claims of a manufacturer.

Surfaces with a natural finish offer different cleaning problems than do painted surfaces. They should be cleaned only according to the manufacturer's instructions.

Existing metal and adjacent surfaces should be kept clean of unwanted paint, dirt, grease, dust, efflorescence, salts, and stains. The cleaning materials used should not harm the metal, its finish, or adjacent surfaces. Abrasives and caustic or acid cleaning agents should not be used. The finish producer's recommendations should be followed in every case.

Adjacent surfaces should be cleaned as recommended by their manufacturers or the associations representing the manufacturers. Cleaning should not harm the surfaces in any way.

Refer to Chapter 4 under "Cleaning and Maintaining Existing Metal Surfaces" and Chapter 5 under "Cleaning Soiled Surfaces That Do Not Need Repair" for additional information.

Installing New Fabrications in Existing Spaces

The requirements discussed earlier under "Installing New Fabrications in New Buildings" are equally applicable when a new metal fabrication is to be installed in an existing space. The requirements are similar, of course, because many metal fabrications are installed into completed spaces, even in all-new buildings. There are, however, some additional considerations that must be taken into account when a metal fabrication is installed into an older building, many of which are addressed here.

Since the number of possible metal fabrications is vast, it is not possible to cover the requirements for installing every one of them in one book. The following discussion is, therefore, generic and should be viewed in that light. The suggestions here may not apply to every individual metal fabrication, but the principles should be applicable to most. In the end there is no substitute for professional judgment, the careful application of which is imperative to the appropriate use of the following data.

In addition to the suggestions that follow, some of the discussions of specific metal fabrications in Chapter 6 include requirements related to installing them in existing buildings.

Preparation

Just as preparations are necessary before a new metal fabrication can be installed in a new space, it is necessary to make preparations before installing a new metal fabrication in an existing space. In the latter case, however, the preparations are likely to be more extensive.

Before work is begun, adjacent spaces that are to remain unaltered should be completely secured and rendered dustproof. Furnishings and equipment there should be removed or adequately protected from damage during construction operations; and fire separations required by code or other legal restraints should be in place.

Submittals.   The types of submittals required when a utilitarian metal fabrication is to be installed in an existing building include shop and setting drawings, manufacturers' literature, cutting and drilling drawings, samples, and narrative descriptions. Refer to "Control," under "Design and Control" in Chapter 6, for detailed discussions about submittals.

Conditions at the Site.   As is true when both the fabrication and the space are new, the actual site where a new metal fabrication is to be installed in an existing building should be examined carefully by the installer. When the fabrication will be installed in an existing space, performing two examinations is usually best. The first should be made before demolition work begins. A report of this examination should be sent to the designer and the owner so that modifications in the proposed demolition can be made, if necessary, to accommodate the fabrication's design.

The second examination should be made after demolition in the space has been completed. This examination is necessary to verify that existing metal fabrications and other construction to which the new fabrication is to be attached, and the areas and conditions where it will be installed, are in proper satisfactory condition to receive the new fabrication. Preparatory installation work should not be started and the metal fabrication should not be installed until unsatisfactory conditions have been corrected.

Protection of Persons and Property.   It is usually best not to begin any kind of demolition or construction work in any portion of an existing building until that part of the building has been vacated by the owner, its use has been discontinued, and the owner has performed the necessary preparatory work in that space. It is not only difficult for the owner's employees to work in a portion of a building where construction is under way—it can sometimes be dangerous. Nevertheless, minor construction projects are often carried out in occupied spaces.

To avoid interfering with the owner's use of the premises, before the organization doing the work related to metal fabrication begins operations, it should be required to notify the owner and authorities owning or controlling wires, conduits, pipes, and other services affected by the work. When disconnecting pipes or services is necessary, this work should be done in accordance with the requirements of the owner and the company, utility, or local authority having jurisdiction. Disconnection should be done only at the convenience of the owner.

An organization working in an existing building is usually made solely responsible for the safety of persons and property affected by its work. The organization should thus provide temporary barricades, fences, shoring, warning lights, dust barriers and partitions, casing of openings, rubbish chutes, temporary closures, and other, similar means of protection. Further arrangements may be made, of course, with the agreement of all parties concerned. Regardless of who assumes responsibility, all agreements should be drafted carefully and signed by all parties concerned. Using standard agreement forms drafted by such organizations as the American Institute of Architects (AIA) will better ensure that everyone has a clear understanding of each party's responsibilities.

When working close to finished existing vegetation, paving, finished interior and exterior surfaces, casework, cabinetwork, equipment, accessories, and devices that remain in place while work is being done, care should be taken to protect those items from staining and other harm. Where items and surfaces that are to be a part of the building after the work has been completed are temporarily removed to perform the work, they should also be protected from damage. Damage done during the work should be satisfactorily repaired by the party causing the damage. Stains should be completely removed, without damaging the stained surfaces. Damaged or stained items that cannot be satisfactorily repaired or cleaned should be removed and replaced with new matching items, at no cost to the owner.

Injury to adjacent facilities and people must be prevented. The safe passage of people around areas where demolition or construction work is taking place must be assured. Temporary covered passageways may be needed, or even required, by the authorities having jurisdiction. Shoring, bracing, or other supports should be provided to prevent movement, settlement, or collapse of facilities that are to remain in place.

Adequate temporary waterproof protection should be provided when work related to a metal fabrication, including the need for access to the location where it will be installed, requires removing parts of existing walls, roofs, or other parts of a building's shell. The protection should be left in place at all times, except when the openings are actually in use, until the existing construction has been permanently replaced and has provided a watertight seal. Parts of an existing building should never be removed without the owner's specific written agreement.

Conduits, drains, pipes, wires, ducts, and mechanical equipment remaining in use and subject to damage during construction activities should be protected and supported.

Airtight covers should be provided for return air and exhaust grilles and duct openings in areas where work will occur, to prevent dirt and dust from entering the air conditioning or ventilating systems. Alternatively, return air and exhaust grilles and duct openings may be capped with a filter to prevent the entry of dust and dirt.

Finish floor coverings and other interior finishes and the owner's furnishings and equipment should be protected from marring and other damage. The protection should be maintained and left in place until the surface, furnishing, or equipment being protected is no longer subject to damage from construction operations.

Heavy fire-retardant building paper should be used to protect existing floor and roof areas. Paths that will carry wheeled traffic, including access ways, should be covered with heavy boards or plywood.

Where portions of walls, partitions, or other vertical construction, or backup for exterior walls are to be removed, the floor on both sides of the construction being demolished should be covered to a distance of at least four feet or to the adjacent wall, whichever is shorter. Where ceilings are to be removed, the entire floor surface should be covered.

Removal and Disposition of Existing Construction.   When new metal fabrications are to be installed in an existing building, at least some selective demolition, dismantling, cutting, patching, and removal of existing construction is usually necessary.

The demolition and removal of existing construction should be done carefully, with the minimum possible interference with existing building or site operations, inconvenience to occupants and the public, danger to persons, and damage to existing materials, furnishings, and equipment that are to remain in place.

When a building is to remain even partly occupied, demolition and removals should be done as quietly as is practicable. To prevent long-term shutdown of the area where the work is being done, demolition and removal should be done with deliberate speed once begun.

Every item or material that will interfere with the installation of new fabrications should be removed. When possible, removed items should be salvaged and reinstalled in their original location. The methods used to remove existing materials should be those required to complete the work, within the limitations of governing regulations. The work should then proceed systematically. Walls and partitions should be demolished in small sections, particularly if they are built of concrete or masonry. Materials should be removed and stored so as not to impose excessive loads on supporting walls, floors, or framing. Removed materials that must be taken away from the building should be lowered to the ground. The use of chutes and the direct dropping of materials should not be permitted, because of the potential harm such methods can do to portions of the existing building that are not being removed.

Temporary struts, bracing, or shoring should be installed and left in place to avoid collapse until the new work provides adequate bracing and support.

Existing materials and equipment that are to remain in place should be protected and, if damaged, repaired promptly and properly. When the damage is too severe for proper repair, the damaged item should be removed and a new one installed in its place.

Items and materials that must be removed to allow work to be done must be removed carefully. Those that are to be reinstalled should be protected and stored carefully in a safe location until they can be reinstalled.

Some items that must be removed will not be reinstalled. These fall into one of three categories, the first being all items that are to remain the owner's property. These should be removed carefully, placed in a safe place designated by the owner, and protected.

In the second category are items that have some value but the owner has not chosen to keep them. Such items, which are sometimes called contractor salvage, usually become the property of the contractor to do with as he wishes. They should be promptly removed from the owner's property so that they do not accumulate and clutter the property or become a hazard. An owner will occasionally permit a contractor to sell such materials or items from the owner's property, but such sale is usually prohibited.

Every demolition project produces material in the third category, which includes debris and rubbish and other materials that are not to be reused, become the property of the owner, left in place, kept and reused in the project, or suitable for contractor salvage. These materials ordinarily become the property of the contractor. They should be promptly removed from the owner's property so that they will not constitute a hazard or a liability problem for the owner. An owner will sometimes permit a contractor to burn, bury, or otherwise dispose of such materials or items on the owner's property, but doing so is more often prohibited.

Demolition and removal of debris should interfere minimally with roads, streets, alleys, walks, and other adjacent facilities. Such facilities should not be closed or obstructed without permission from the authorities having jurisdiction. Alternative routes should be provided around closed or obstructed trafficways if governing regulations so require.

The ownership of removed, reusable, or salable property can often become a source of contention. It is usually best from the owner's point of view to state in whatever agreements are drawn up for the work that every item except rubbish and debris belongs to the owner until he formally refuses to accept it. Then, there can be no dispute about who owns removed property. Such an arrangement will sometimes add to the cost of a project, however, because the contractor must assume that there will be no salvageable property accruing to him and will not make an allowance for it in his price.

One way to help head off disputes about the ownership of removed materials is for the owner to remove commonly disputed objects before the contractor is given access to the work area. The owner would not need to remove items that were to be reused in the project, only those that might otherwise be discarded or become the property of the contractor but which the owner wants to keep. Such items might include locks, lock or latch sets, door closers, sink and lavatory fittings, laboratory equipment, furnishings, and portable equipment.

It is often necessary to cut or drill existing construction to install new metal fabrications in an existing space. Whenever possible, such cutting should be done by hand or using small power tools. Holes and slots should be cut neatly to the size required, with minimal disturbance of adjacent materials. Round holes in concrete slabs, floors, and walls should be made with core drills of the required sizes and types. Square and rectangular holes should be made by line drilling, then using chipping hammers to remove the material between the drill holes. The use of large air hammers should not be permitted in an existing building, because they have great potential for causing harm to adjacent parts of the building. For similar reasons, air compressors should not be operated inside a building.

No drilling or cutting of columns, beams, joists, girders, ties, or other structural supporting elements should be permitted unless the owner specifically approves it in each case or doing so is indicated on structural drawings prepared by a professional engineer. The owner should issue permission to cut or drill structural elements of a building only after consulting with a structural engineer. Cutting existing concrete slab reinforcement should be avoided.

In order to protect the interior of a building and its contents, openings in exterior walls and the roof should be covered temporarily when not in use, and patched as soon as the new work has been installed.

Demolition may uncover products that appear to contain friable asbestos or other hazardous materials. If so, all work should immediately be halted until the extent of the hazard has been determined and safe, suitable methods developed to remove it. Particular care should be taken to protect workers, building occupants, and the public from hazards associated with such materials and to comply with governmental laws and regulations. Discussion of the methods that should be used to remove or encapsulate asbestos and other hazardous materials is beyond the scope of this book.

Attachments and Anchors. When the supports for a new metal fabrication are already in place, the attachments and anchors may not be of the same types that would be used if both the fabrication and its supporting structure were new. Refer to "Installation of Metal Fabrications" below for a discussion of the types of anchorage and anchor devices that might be used. When the supporting structure is already in place it is usually not necessary to supply inserts and anchorage devices before the fabrications arrive at the site.

If, however, any part of the supporting structure will be newly placed specifically to support or anchor the new fabrication, then such items as sleeves, concrete inserts, anchor bolts, and miscellaneous items with integral anchors to be embedded in concrete or masonry must be delivered to the site earlier to be built into the other construction.

Installation of Metal Fabrications

The installation of metal fabrications must be coordinated with the work of other trades. Fabrications must be delivered and installed at the proper time.

Should conditions be encountered that differ significantly from those expected, the person discovering the difference should immediately notify other parties concerned, especially the owner, so that alternative methods can be worked out for handling the situation.

Damaged metal fabrications should not be installed, because it is generally much easier to repair a damaged fabrication before it has been installed. Often, damage will not be repairable after a fabrication has been installed, and the fabrication must be removed to permit repair. Damaged metal items that have been installed and cannot be repaired in place should be removed and repaired or new, undamaged units installed. Installing already damaged metal fabrications is one of the greatest causes of later problems with them.

Metal fabrications should be installed in accordance with the manufacturer's latest published instructions and the approved shop drawings and be securely anchored in place. There will be minor differences in the installation methods available, depending on whether a unit is to be installed before its space is built, as the space is erected, or in an already prepared space. The first two situations occur when part of an existing construction must be removed to permit installation of a new metal fabrication, or for other reasons.

There are three major advantages to installing a unit before its space is built or while its space is being erected. First, this makes it possible to build anchoring and support devices, such as anchors and inserts, into new concrete, masonry, and stone while it is being erected. Second, this permits shop fabrication and delivery of metal units in large sections, often in one piece. Finally, this makes extreme accuracy in fabricating the metal units somewhat less important, because the surrounding construction can be adjusted slightly during construction. However, this can be a disadvantage if it tempts the fabricator to give too little attention to accuracy. The result can be aesthetically less than pleasing.

Even in new buildings, metal fabrications are often installed after the surrounding construction has been erected. This condition is the norm when a fabrication is to be installed in an existing structure. Sometimes, though, it makes more sense to remove part of the existing construction to install a metal fabrication—and sometimes this removal is necessary. It is hard to imagine, for example, trying to install a new steel stair in an existing building without considerably demolishing the existing construction.

The types of anchors used may vary, of course, depending on when in the construction process the fabrications are installed. A metal fabrication to be anchored to concrete or masonry that has not yet been constructed can be anchored using threaded fasteners into inserts or to fasteners that have been set into masonry or stone joints. If the masonry, stone, or concrete are already there, then toggle bolts, through bolts, threaded fasteners into expansion shields in drilled holes, and similar fasteners and anchors will be necessary.

As a general rule, fastening and anchoring are more complicated when a metal fabrication is to be installed in an existing building than when all the construction is new. This rule holds true regardless of the type of construction. Even wood studs, whose location can easily be marked in new construction so that the marks are visible after the finish has been applied, must be located individually in an existing building. Anchoring plates can easily be added to metal stud walls and partitions before the finish is applied, but when the plates have not been installed and the finish is already in place, anchoring becomes more difficult. Welding anchor bolts to new structural steel to anchor metal fabrications is relatively easy. But old steel will have been painted or maybe even covered with fireproofing materials, which must then be removed before an anchor can be attached, and be put back afterward.

In order to install metal fabrications it is often necessary to perform cutting, drilling, and fitting of either the fabrication or the adjacent construction. Such work should be done so that each fabrication can be set accurately in the proper location, alignment, and elevation. Fabrications should have their surfaces and edges plumb, level, true, and free of rack, as measured from established lines and levels.

Exposed connections should fit accurately together to form tight, hairline joints. The best field connections of pipes and tubes are made using internal sleeves. Connections that are not to be left as open joints or screwed, bolted, or riveted together should be welded. Where possible, they should be welded in the shop. If shipping-size limitations preclude shop welding, the joints should be welded in the field. In either case, exposed welds should be ground smooth. Paint coats should then be touched up, using the same paint that was used for the shop coat. The surfaces of exterior units that have been hot-dip galvanized after fabrication and that will have bolted or screwed field connections or been factory prefinished should not be welded, cut, or abraded. Refer to Chapter 6 for additional suggestions about welding.

Patching Adjacent Materials and Surfaces

Installing a new metal fabrication in an existing space will almost always be accompanied by the need to patch or repair adjacent surfaces. Sometimes it will have been necessary to remove adjacent materials or items, which must then be reinstalled.

Removed materials and items that are to be reused should be handled carefully to prevent damage and stored in a safe location until reinstalled. Anything to be reused that has become damaged should be repaired or discarded and new equal products provided. Salvaged items to be reinstalled should be cleaned and put in proper working order before being reinstalled. Reused materials should be in good condition, without objectionable chips, cracks, splits, checks, dents, scratches, or other defects. Operating items should work properly. Missing parts necessary to complete each installation should be provided.

When possible, the materials and items removed to permit installation of new metal fabrications in existing spaces should be reinstalled in the same location from which they were removed. Sometimes, though, a contractor will find it less costly to discard removed materials and install new ones in their place. This activity should not be discouraged without good reason.

When removed materials or items are damaged beyond repair or are otherwise unsatisfactory for reuse, new materials or items should be substituted. The new materials or items should exactly match the removed ones in every particular, unless the owner agrees that a different appearance or item is satisfactory.

Floor, wall, and ceiling finishes that have been damaged or defaced by cutting, patching, demolition, alteration, or repair work should be restored to a condition equal to that which existed before the work started.

Damaged or unfinished surfaces or materials that are exposed by an alteration, repair, or removal should be repaired and either finished or refinished, as appropriate. If not, the damaged or unfinished surfaces or materials should be removed and new, acceptable, matching surfaces or materials provided. Alternatively, acceptable salvaged materials may be used to make continuous areas and surfaces uniform.

The restoration and refinishing of existing damaged surfaces and items should be done in accordance with the highest standards of the trade involved and in accordance with approved industry standards, as required to match the existing surface. The workmanship used to repair existing materials should conform to that in or adjacent to the space where the alterations are to be made.

Holes and openings in existing floor, wall, and ceiling surfaces resulting from alteration work should be closed and filled to match adjacent undisturbed surfaces.

Cleaning and Protection of Installed Metal Fabrications

The same principles apply to cleaning and protecting new metal fabrications installed in existing spaces as to installing new fabrications in new spaces. Refer to "Cleaning Installed Metal Fabrications" under "Installing New Fabrications in New Buildings" earlier in this chapter for a discussion of this subject.

Where to Get More Information

The AIA Service Corporation's *Masterspec* sections noted in the Bibliography with a [7] contain data related to installing items here called metal fabrications. Unfortunately, they contain few references to solving failure problems. They are, however, excellent references about how the items should have been installed in the first place. This becomes even more important when one realizes that failures often occur because someone installed the item improperly in the first place.

Architectural Graphic Standards contains excellent details showing how metal fabrications should go together. There is unfortunately little in it to help with troubleshooting such fabrications.

John G. Waite's article "Caring for Decorative Metalwork" includes excellent advice about many problems common in decorative wrought iron, cast iron, and zinc items, such as railings, steps, window sills and lintels, and fences.

Excellent information about sealants is to be found in the following publications:

- *Masterspec,* Section 07900.
- *Architectural Graphic Standards.*

- U.S. Department of the Navy Guide Specifications Section 07920, "Sealants and Caulkings."
- U.S. Departments of the Army and Air Force, *Building Construction Materials and Practices: Caulking and Sealing.*
- Sealant Engineering and Associated Lines. "SEAL Guide Specification for Sealants and Caulking."

In addition, most nationally recognized sealant manufacturers offer excellent guidance on sealant selection and installation.

Refer also to other entries in the Bibliography that are followed by a [7].

APPENDIX

Data Sources

NOTE: The following list includes sources of data referenced in the text, included in the Bibliography, or both. **HP** following a source indicates that that source contains data of interest to those concerned with historic preservation.

Adhesive and Sealant Council (ASC)
1500 Wilson Boulevard, Suite 515
Arlington, VA 22209
(703) 841-1112

Advisory Council on Historic
    Preservation
1100 Pennsylvania Avenue, Suite 809
Washington, DC 20004
(202) 786-0503 **HP**

AIA Service Corporation
American Institute of Architects
1735 New York Avenue, N.W.
Washington, DC 20006
(202) 626-7300

Air Movement and Control
    Association
30 West University Drive
Arlington Heights, IL 60004
(312) 394-0150

Allied Building Metals Industries
211 East 43rd Street, #804
New York, NY 10017
(212) 697-5551

Aluminum Association (AA)
818 Connecticut Avenue, N.W.
Washington, DC 20006
(202) 862-5100

Aluminum Extruders Council
4300-L Lincoln Avenue
Rolling Meadows, IL 60008
(312) 359-8160

American Architectural
  Manufacturers Association
  (AAMA)
2700 River Road, Suite 118
Des Plaines, IL 60018
(312) 699-7310

American Association of State and
  Local History
172 Second Avenue North, Suite 102
Nashville, TN 37201
(615) 255-2971 **HP**

American Concrete Institute
P.O. Box 19150, Redford Station
2400 West Seven Mile Road
Detroit, MI 48219
(313) 532-2600

American Galvanizers Association
1101 Connecticut Avenue, N.W.,
  Suite 700
Washington, DC 20036
(202) 857-1119

American Institute of Architects
1735 New York Avenue, N.W.
Washington, DC 20006
(202) 626-7300

American Institute of Architects
Committee on Historic Resources
1735 New York Avenue, N.W.
Washington, DC 20006
(202) 626-7300 **HP**

American Institute of Steel
  Construction
400 North Michigan Avenue
Chicago, IL 60611-4185
(312) 670-2400

American Institute of Timber
  Construction
333 West Hampden Avenue
Englewood, CO 80110
(303) 761-3212

American Iron and Steel Institute
1133 15th Street, Suite 300
Washington, DC 20005
(202) 452-7100

American National Standards
  Institute (ANSI)
1430 Broadway
New York, NY 10018
(212) 354-3300

American Painting Contractor
American Paint Journal Company
2911 Washington Avenue
St. Louis, MO 63103

American Plywood Association
  (APA)
P.O. Box 11700
Tacoma, WA 98411
(206) 565-6600

American Society for Metals
Metals Park, OH 44073
(216) 338-5151

American Welding Society
550 N.W. LeJeune Road
Box 351040
Miami, FL 33135
(305) 443-9353

American Wire Producers
  Association
2000 Maple Hill Street
Yorktown Heights, NY 10598
(914) 962-9052

American Wood Preservers
  Association
P.O. Box 5283
Springfield, VA 21666
(703) 339-6660

American Wood Preservers Bureau
P.O. Box 6058
2772 South Randolf St.
Arlington, VA 22206
(703) 931-8180

Architectural Anodizers Council
1227 West Wrightwood Avenue
Chicago, IL 60614
(312) 871-2550

Architectural Technology
American Institute of Architects
1735 New York Avenue, N.W.
Washington, DC 20006
(202) 626-7300

Architecture
American Institute of Architects
1735 New York Avenue, N.W.
Washington, DC 20006
(202) 626-7300

Associated Builders and Contractors
729 15th Street, N.W.
Washington, DC 20005
(202) 637-8800

Associated General Contractors of
    America
1957 E Street, N.W.
Washington, DC 20006
(202) 393-2040

Association for Preservation
    Technology
Box 2487, Station D
Ottawa, Ont. K1P 5W6, Canada
(613) 238-1972 **HP**

ASTM
1916 Race Street
Philadelphia, PA 19103-1187
(215) 299-5585

Brick Institute of America
11490 Commerce Park Drive,
    Suite 300
Reston, VA 22091
(703) 620-0010

Building Design and Construction
Cahners Plaza
1350 East Touhy Avenue
P.O. Box 5080
Des Plaines, IL 60018
(312) 635-8800

Building Owners and Managers
    Association International
1250 I Street, N.W., Suite 200
Washington, DC 20005
(202) 289-7000

Campbell Center for Historic
    Preservation Studies
P.O. Box 66
Mount Carroll, IL 61053
(815) 244-1173 **HP**

Chain Link Fence Manufacturers
    Institute
1776 Massachusetts Avenue, N.W.,
    Suite 500
Washington, DC 20036
(202) 659-3537

Cold Finished Steel Bar Institute
1120 Vermont Avenue, N.W.
Washington, DC 20005
(202) 857-0350

Color Association of the United
    States
343 Lexington Avenue
New York, NY 10016
(212) 683-9531

*Commercial Remodeling. See
    Commercial Renovation*

Commercial Renovation (formerly
    Commercial Remodeling)
QR, Inc.
1209 Dundee Avenue, Suite 8
Elgin, IL 60120

Commercial Standard (U.S.
    Department of Commerce)
Government Printing Office
Washington, DC 20402
(202) 377-2000

The Construction Specifier
Construction Specifications Institute
601 Madison Street
Alexandria, VA 22314-1791
(703) 684-0300

Copper and Brass Fabricators
Council
1050 17th Street, N.W., #440
Washington, DC 20036
(202) 833-8575

Copper Development Association
(CDA)
Greenwich Office Park 2
Box 1840
Greenwich, CT 06836
(203) 625-8210

Custom Roll Forming Institute
522 Westgate Tower
Cleveland, OH 44116
(216) 333-8848

Decorating Products World
Elkin Mittelman (Publishers)
7335 Topango Canyon Boulevard
Canoga Park, CA 91303
(818) 710-1066

Edgell Communications
7500 Old Oak Boulevard
Cleveland, OH 44130

Environmental Protection Agency
401 M Street, S.W.
Washington, DC 20460
(202) 829-3535

Exteriors
1 East 1st Street
Duluth, MN 55802
(218) 723-9200

Factory Mutual System
1151 Boston–Providence Turnpike
Norwood, MA 02062
(617) 762-4300

Federal Housing Administration
(U.S. Department of Housing and
Urban Development)
451 7th Street, S.W.
Washington, DC 20201
(202) 755-5995

Federal Specifications (U.S. General
Services Administration)
Specifications Unit
7th and D Streets, S.W.
Washington, DC 20407
(202) 472-2205

Federation of Societies for Coatings
Technology
1315 Walnut Street
Philadelphia, PA 19107
(215) 545-1506

Ferroalloys Association
1511 K Street, N.W., #833
Washington, DC 20005
(202) 393-0555

Forest Products Laboratory
U.S. Department of Agriculture
Gifford Pinchot Drive
P.O. Box 5130
Madison, WI 53705
(698) 264-5600

General Services Administration
General Services Building
18th and F Streets, N.W.
Washington, DC 20405
(202) 655-4000

Gypsum Association
1603 Orrington Avenue
Evanston, IL 60201
(312) 491-1744

Hartford Architecture Conservancy
130 Washington Street
Hartford, CT 06106
(203) 525-0279 **HP**

Heritage Canada Foundation
Box 1358, Station B
Ottawa, Ont. K1P 5R4, Canada
(613) 237-1867 **HP**

Historic Preservation. See
National Trust for Historic
Preservation. **HP**

Illinois Historic Preservation Agency
Division of Preservation Services
Old State Capitol
Springfield, IL 62701
(217) 782-4836 **HP**

International Institute for Lath and
  Plaster
c/o W. F. Pruter Associates
3127 Los Feliz Boulevard
Los Angeles, CA 90039
(213) 660-4644

International Lead Zinc Research
  Organization
P.O. Box 12036
Research Triangle Park, NC 27709-
  2036
(919) 361-4647

Institute for Applied Technology/
  Center for Building Technology
National Bureau of Standards
U.S. Department of Commerce
Washington, DC 20540
(202) 342-2241

International Masonry Institute
823 15th Street, N.W., #1001
Washington, DC 20005
(202) 783-3908

Iron Casting Society
455 State Street
Cast Metals Federation Building
Des Plaines, IL 60016
(312) 299-9160

*Journal of Special Coatings and
  Linings. See* Contact Steel
  Structures Painting Council.

Lead Industries Association
292 Madison Avenue
New York, NY 10017
(212) 578-4750

Library of Congress
1st Street, N.E.
Washington, DC 20540
(202) 287-5000 **HP**

McGraw-Hill Book Company
1221 Avenue of the Americas
New York, NY 10020
(212) 997-2271

Metal Architecture
7450 North Skokie Boulevard
Skokie, IL 60077
(312) 674-2200

Metal Building Component
  Manufacturers Association
1133 15th Street, N.W.
Washington, DC 20005
(202) 429-9440

Metal Building Manufacturers
  Association
1230 Keith Building
Cleveland, OH 44115
(216) 241-7333

Metal Construction Association
1133 15th Street, N.W., Suite 1000
Washington, DC 20005
(202) 429-9440

Metal Lath/Steel Framing
  Association
600 South Federal Street, Suite 400
Chicago, IL 60605
(312) 922-6222

Modern Metals
Delta Communications, Inc.
400 North Michigan Avenue
Chicago, IL 60611

National Alliance of Preservation
  Commissions
Hall of the States
444 North Capitol Street, N.W.,
  Suite 332
Washington, DC 20001
(202) 624-5490 **HP**

National Association of Architectural
  Metal Manufacturers (NAAMM)
600 South Federal Street, Suite 400
Chicago, IL 60605
(312) 922-6222

National Association of Corrosion
    Engineers
1440 South Creek
Houston, TX 77084
(713) 492-0535

National Bureau of Standards (NBS).
    See National Institute of
    Standards and Technology.

National Coil Coaters Association
1900 Arch Street
Philadelphia, PA 19103
(215) 564-3484

National Concrete Masonry
    Association
P.O. Box 781
Herndon, VA 22070
(703) 435-4900

Natonal Decorating Products
    Association
1050 North Lindbergh Boulevard
St. Louis, MO 63132
(314) 991-3470

National Fire Protecton Association
Batterymarch Park
Quincy, MA 02169
(800) 344-3555

Natonal Forest Products Association
1619 Massachusetts Avenue, N.W.
Washington, DC 20036
(202) 797-5800

National Institute of Standards and
    Technology (formerly National
    Bureau of Standards)
Gaithersburg, MD 20234
(301) 975-2000

National Institute of Standards and
    Technology (formerly National
    Bureau of Standards)
Center for Building Technology
Gaithersburg, MD 20234
(301) 975-5900

National Ornamental and
    Miscellaneous Metals Association
2996 Grandview Avenue, N.E., #109
Atlanta, GA 30305
(404) 237-5334

National Paint and Coatings
    Association
1500 Rhode Island Avenue, N.W.
Washington, DC 20005
(202) 462-6272

National Paint, Varnish and Lacquer
    Association
(now the National Paint and Coatings
    Association)

National Preservation Institute
P.O. Box 1702
Alexandria, VA 22313
(703) 393-0038 **HP**

National Trust for Historic
    Preservation
1785 Massachusetts Avenue, N.W.
Washington, DC 20036
(202) 673-4000 **HP**

Occupational Safety and Health Ad-
    ministration (U.S. Department of
    Labor)
Government Printing Office
Washington, DC 20402
(202) 783-3238

The Old-House Journal
435 Ninth St.
Brooklyn, NY 11215
(718) 788-1700 **HP**

Painting and Decorating Contractors
    of America
3913 Old Lee Highway, Suite 33B
Fairfax, VA 22030
(703) 359-0826

See also PWC magazine.

Painting and Decorating Contractors
of America
Washington State Council
27606 Pacific Highway South
Kent, WA 98032
(206) 941-8823

Porcelain Enamel Institute
1101 Connecticut Avenue, N.W.,
Suite 700
Washington, DC 20036
(202) 857-1134

The Preservation Press
National Trust for Historic
Preservation
1785 Massachusetts Avenue, N.W.
Washington, DC 20036
(202) 673-4000 **HP**

Preservation Resource Group
5619 Southampton Drive
Springfield, VA 22151
(703) 323-1407 **HP**

Product Standards of NBS (U.S. Department of Commerce)
Government Printing Office
Washington, DC 20402
(202) 783-3238

*PWC Magazine: Painting and
Wallcovering Contractor*
(Magazine of Painting and Decorating
Contractors of America)
130 West Lockwood Street
St. Louis, MO 63119
(314) 961-6644

Sealant Engineering and Associated
Lines (SEAL)
P.O. Box 24302
San Diego, CA 92124
(619) 569-7906

Sealant and Waterproofing Institute
3101 Broadway, #300
Kansas City, MO 64111-2416
(816) 561-8230

Sheet Metal and Air Conditioning
Contractors National Association
(SMACNA)
8224 Old Courthouse Road
Vienna, VA 22180
(703) 790-9890
For publications contact:
SMACNA Publications Department
P.O. Box 70
Merrifield, VA 22116
(703) 790-9890

Society for the Preservation of New
England Antiquities
141 Cambridge Street
Boston, MA 02114
(617) 227-3956 **HP**

Southern Pine Inspection Bureau
4709 Scenic Highway
Pensacola, FL 32504
(904) 434-2611

Steel Structures Painting Council
4400 Fifth Avenue
Pittsburgh, PA 15213
(412) 578-3327

Technical Preservation Services
U.S. Department of the Interior,
Preservation Assistance Division
National Park Service
P.O. Box 37127
Washington, DC 20013-7127
(202) 343-7394 **HP**

Truss Plate Institute
583 D'Onofrio Drive, Suite 200
Madison, WI 53719
(608) 833-5900

Underwriters Laboratories
333 Pfingsten Road
Northbrook, IL 60062
(312) 272-8800

U.S. Department of Agriculture
14th Street and Independence
    Avenue, S.W.
Washington, DC 20250
(202) 447-4929

U.S. Department of Commerce
14th Street and Constitution
    Avenue, N.W.
Washington, DC 20230
(202) 377-2000

U.S. Department of Commerce for
    PS (Product Standard of NBS)
Government Printing Office
Washington, DC 20402
(202) 783-3238

U.S. Department of the Interior
National Park Service
P.O. Box 37127
Washington, DC 20013-7127
(202) 343-7394 **HP**

U.S. General Services Administration
Historic Preservation Office
18th and F Streets, N.W.
Washington, DC 20405
(202) 655-4000 **HP**

U.S. General Services Administration
Specifications Unit
7th and D Streets, S.W.
Washington, DC 20407
(202) 472-2205/2140

U.S. Government Printing Office
North Capitol and H Streets, N.W.
Washington, DC 20405
(202) 275-3204

United States Gypsum Company
101 South Wacker Drive
Chicago, IL 60606
(312) 606-4000

Van Nostrand Reinhold
115 Fifth Avenue
New York, NY 10003
(212) 254-3232

Welded Steel Tube Institute
522 Westgate Tower
Cleveland, OH 44416
(216) 333-4550

West Coast Lumber Inspection
    Bureau
P.O. Box 23145
Portland, OR 97223
(503) 639-0651

Western Wood Products Association
1500 Yeon Building
Portland, OR 97204
(503) 224-3930

John Wiley & Sons
605 Third Avenue
New York, NY 10158
(212) 850-6000

Zinc Institute
292 Madison Avenue
New York, NY 10017
(212) 578-4750

Glossary

Terms that have been set in SMALL CAPS are defined elsewhere in this Glossary.

AIR DRYING. When used in conjunction with ORGANIC COATINGS, air drying means that a COATING will cure properly without the use of forced air. PAINT, for example, is always air drying.

ALLIGATORING. Cracks in an applied coating that resemble an alligator's hide. A more severe form of CHECKING.

ALLOY. The material formed when two or more metals, or metals and nonmetallic substances, are joined by being dissolved into one another while molten. There are four common groups of alloys: (1) two-layer alloys, in which the materials are mutually insoluble and thus remain separately distinguishable in the alloy; (2) solid solutions, which are a homogeneous mass of crystals; includes brass, bronze, Monel metal, and austenite; (3) and (4) EUTECTIC and EUTECTOID, analogous groups. Mixtures that have different freezing temperatures than either of their constituents. Includes tin solder (made from tin and lead) and iron/carbon alloys, such as pig iron, wrought iron, cast iron, malleable iron, and steel.

ANNEAL. To heat, then cool at a controlled rate. Makes metals tougher (that is, less BRITTLE) and stronger.

AS FABRICATED FINISH. A finish imparted by the forming of aluminum or copper-alloy items. Synonymous with MILL FINISH as applied to ferric metals. Does not mean the finish after fabrication of the metal into a usable part or assembly.

BAKING FINISH. An organic coating formulated or modified to cure only by force drying or baking.

BINDER. *See* FILM FORMER.

BLISTER. Bubbles, small or large, in an ORGANIC COATING. May be caused by entrapped gases, separation between coats, or of the coating from the substrate for some other reason.

BRITTLE. Lacking in TOUGHNESS; nonductile. *See also* DUCTILE.

CARBURIZED METAL. Metal impregnated with carbon.

CEMENTATION. Forming an alloy between a coating metal and the coated metal. Often accomplished by covering the coated metal with a powder of the coating metal, then applying heat until the two metals combine (form an ALLOY).

CHECKING. Cracks in an applied coating that resemble ALLIGATORING but are much smaller and less noticeable. In its early stages also called CRAZING, crowfooting, or hairlining.

CLEAR ORGANIC COATING. An ORGANIC COATING applied as a liquid that leaves a transparent film on the coated surface; either AIR DRYING or a BAKING FINISH. Some air-drying clear organic coatings may be applied in the field and thus fit the definition for PAINT but are not considered paint in the industry.

COATING. A material used to coat a metal, either to protect it or decorate it. Included are metallic coatings, PAINT, ORGANIC COATINGS other than paint, anodic coatings, vitreous coatings, and laminated coatings. All parts of a COATING SYSTEM are called a coating, whether collectively or individually.

COATING SYSTEM. Some coatings consist of several layers or coats, called collectively a coating system. Each material in a coating system is called a COATING.

CRAZING. Fine cracks in an applied coating. In their early stages, CHECKING and ALLIGATORING are sometimes called crazing.

CRITICAL WORKING TEMPERATURE. The temperature at which a metal can reorganize its molecular structure during working and at which it will automatically recover its plasticity after deformation.

DEGREASING. Removing oil and grease from metal.

DESCALING. Removing mill scale from metals.

DUCTILE. Capable of being drawn out into another shape or hammered or rolled thin, without fracturing or returning to its original shape when the applied force is removed.

ELASTIC. Capable of returning to its original shape after being deformed. Metals are elastic to varying degrees. The elasticity of a given material may vary with its temperature.

ELASTIC DEFORMATION. Deformation caused by a STRESS small enough so that when the stress is removed the metal returns to its original shape. *See also* PLASTIC DEFORMATION.

ELASTIC LIMIT. The amount of STRESS that will cause a given material to deform or set permanently. *See also* MODULUS OF ELASTICITY.

ENAMEL. Older enamels were essentially just pigmented varnishes. To-day's enamels are AIR DRYING or BAKING FINISH ORGANIC COATINGS that are virtually indistinguishable from PAINT, except for an additive or a difference in formulation in the baking-finish types so that they require adding heat to make them cure properly.

EUTECTIC ALLOY. A mechanical mixture with a lower freezing point than any of the metals used to make it. See also ALLOY.

EUTECTOID ALLOY. Similar to EUTECTIC ALLOYS but used when speaking of a SOLID SOLUTION rather than a liquid one, to which the term "eutectic" refers. See also ALLOY.

EXTENDER. A material used to impart some desirable quality to a PAINT or COATING that the paint or coating alone does not have. May make a paint or coating flow more easily. They do not usually increase the hiding ability of a paint or coating.

FAILURE. Every type of defect, from the slight crazing of a COATING to the total collapse of a NONSTRUCTURAL METAL ITEM, and everything in between.

FILM FORMER. A nonvolatile ingredient in a PAINT or COATING that binds its solid particles together. Usually called a binder.

FLAKING. In an applied COATING, an advanced form of BLISTERING, AL-LIGATORING, or other cracking in which small pieces of the coating fall off. See also SCALING.

HARDNESS. In metals, resistance to abrasion, scratching, indentation, and PLASTIC DEFORMATION.

INORGANIC FINISH. Any finish that does not fit into the National As-sociation of Architectural Metal Manufacturers (NAAMM) classification of "ORGANIC COATINGS."

LACQUER. A transparent, AIR DRYING or BAKING FINISH ORGANIC COATING. Often but not always based on a modified cellulose resin.

MALLEABLE. Capable of being hammered or rolled into a thinner shape without fracture.

METAL FABRICATION. See NONSTRUCTURAL METAL ITEM.

MILL FINISH: On ferrous metal products or aluminum, the finish im-parted at the mill.

MODULUS OF ELASTICITY. The ratio of stresses lower than the ELASTIC LIMIT to strains.

NONSTRUCTURAL METAL ITEM. Any item made from metal that is used in building or related construction but does not form part of a building's structural support system; also called a metal fabrication.

ORGANIC COATING. A group of COATINGS, classified as such by the Na-tional Association of Architectural Metal Manufacturers (NAAMM), that includes AIR DRYING and BAKING FINISH opaque ENAMELS and PAINTS, and clear coatings.

OTHER ORGANIC COATINGS. All ORGANIC COATINGS that do not fit into the definition of PAINT.

PAINT. An AIR DRYING ORGANIC COATING that can be applied either in the shop or the field, but usually in the field. An air-drying ENAMEL is paint by this definition; BAKING FINISH enamel is not. Metal PRIMERS for field-applied paints are part of the PAINT SYSTEM and are therefore also called paint. Primers can be applied by hand or machine in factory or shop and may be either air-drying or baking-finish coatings. Organic coatings that do not fit this definition of paint are called OTHER ORGANIC COATINGS in the text. *See also* CLEAR ORGANIC COATINGS.

PAINT SYSTEM. Collectively, the several coats necessary to produce a complete PAINT COATING. The materials used as the coats in a paint system are called paint. These include PRIMERS, emulsions, ENAMELS, fillers, sealers, stains, and other applied materials used as prime, intermediate, or finish coats.

PICKLING. Immersing a metal in a diluted acid solution to remove mill scale.

PIGMENT. Solid particles, usually as a fine powder, in a PAINT or COATING that provide the material's color and its ability to cover and fill. The ingredient that makes paint and ORGANIC COATINGS opaque.

PLASTIC. Able to remain in a deformed shape after the loads that created the deformation have been removed.

PLASTIC DEFORMATION. Deformation caused by a STRESS exceeding the ELASTIC LIMIT of the material so that it does not return to its original shape when the stress is removed. *See also* ELASTIC DEFORMATION.

PRIMER. That coat of a PAINT or COATING SYSTEM that is applied to bare metal to prepare or prime it to receive succeeding coats. Most bare metal to be field painted is given a prime SHOP COAT.

PROCESS FINISH. Finishes on aluminum or copper alloys, other than AS FABRICATED finishes.

PROFESSIONAL PAINT MATERIALS. PAINT materials specifically formulated for application by professional painters. Often require thinning and the addition of materials not included in the paint as sold.

SCALING. In an applied COATING, an advanced form of FLAKING in which the pieces that fall off are large.

SHOP COAT. A coat of PAINT applied in the shop as a PRIMER for the finish coats of paint to be applied in the field. Protects the metal until final coats are applied.

SLAG. The residue left after the metal is removed from iron ore.

SOLIDS. The nonvolatile ingredients in a paint or organic COATING SYSTEM. Includes the FILM FORMER or binder and PIGMENTS.

SOLID SOLUTION. A metal ALLOY in which the metals are mutually soluble.

Solvent. In a paint or organic coating system's components, a volatile liquid used to dissolve FILM FORMER and PIGMENT. *See also* THINNER.

Standard-thickness plaster. Field-applied plaster that exceeds 1/4 inch in thickness. Excludes skim-coat plaster and veneer plaster.

Strain. The change in cross-sectional area of a body produced by STRESS. Measured in inches per inch of length.

Strength. Ability to withstand stress.

Stress. The intensity of a mechanical force acting on a body: either tensile, compressive, or shear. Stress is measured by dividing total force by the area over which the force acts, as pounds per square inch, for example.

Telegraphing. The appearance of underlying defects or coloration through an applied COATING.

Thinner. A liquid used for thinning, or reducing the viscosity of, a component of a paint or organic coating system. When the component is solvent based, the thinner will be a volatile material. If the component is water based, the thinner is usually water. Often used interchangeably with SOLVENT.

Toughness. Resistance to impact, in metals.

Varnish. A clear organic finish comprised of oils and resins. Oleoresinous varnishes cure by chemical reaction, spirit varnishes by evaporation.

Vehicle. Resins that form a flexible film and bind a pigment together in a PAINT or ORGANIC COATING. Contains the paint or organic coating's FILM FORMER or binder.

Yield point. The level of STRESS at which a relatively great increase in STRAIN occurs without a corresponding increase in stress. Applying stresses in excess of a metal's yield point will usually result in structural damage to it.

Bibliography

Most items listed here are followed by one or more numbers in brackets. Those numbers list the chapters in this book to which that bibliographical entry applies.

The **HP** designation following some entries here indicates that that entry has particular significance for historic preservation projects.

The sources for many of these entries, with addresses and telephone numbers, are listed in the Appendix.

Abate, Kenneth. 1989. Metal Coatings, Fighting the Elements with Superior Paint Systems, Part I. *Metal Architecture*, 5(6)(June): 10, 69. [5].

―――. 1989. Metal Coatings, Fighting the Elements with Superior Paint Systems, Part II. *Metal Architecture*, 5(7)(July): 40–41. [5].

AIA Service Corporation. *Masterspec*, Basic: Section 02831, Chain Link Fencing and Gates, 5/89 ed. The American Institute of Architects. [6, 7].

―――. *Masterspec*, Basic: Section 03310, Concrete Work, 5/87 ed. The American Institute of Architects. [2].

―――. *Masterspec*, Basic: Section 03410, Structural Precast Concrete, 8/87 ed. The American Institute of Architects. [2].

―――. *Masterspec*, Basic: Section 03450, Architectural Precast Concrete, 5/86 ed. The American Institute of Architects. [2].

―――. *Masterspec*, Basic: Section 03470, Tilt-Up Concrete Construction, 5/87 ed. The American Institute of Architects. [2].

―――. *Masterspec*, Basic: Section 04200, Unit Masonry, 5/85 ed. The American Institute of Architects. [2].

―――. *Masterspec*, Basic: Section 04230, Reinforced Unit Masonry, 5/85 ed. The American Insitute of Architects. [2].

———. *Masterspec,* Basic: Section 04400, Stonework, 11/86 ed. The American Institute of Architects. [2].

———. *Masterspec,* Basic: Section 05120, Structural Steel, 8/86 ed. The American Institute of Architects. [2].

———. *Masterspec,* Basic: Section 05210, Steel Joists and Joist Girders, 8/86 ed. The American Institute of Architects. [2].

———. *Masterspec,* Basic: Section 05310, Steel Deck, 5/89 ed. The American Institute of Architects. [2].

———. *Masterspec,* Basic: Section 05400, Cold-Formed Metal Framing, 5/89 ed. The American Institute of Architects. [2].

———. *Masterspec,* Basic: Section 05500, Metal Fabrications, 8/88 ed. The American Institute of Architects. [3, 4, 5, 6, 7].

———. *Masterspec,* Basic: Section 05520, Handrails and Railings, 5/87 ed. The American Institute of Architects. [3, 4, 6, 7].

———. *Masterspec,* Basic, Section 05580, Sheet Metal Fabrications, 5/88 ed. The American Institute of Architects. [3, 6, 7].

———. *Masterspec,* Basic: Section 05700, Ornamental Metalwork, 2/88 ed. The American Institute of Architects. [3, 4, 6, 7].

———. *Masterspec,* Basic: Section 05715, Prefabricated Spiral Stairs, 2/83 ed. The American Institute of Architects. [3, 4, 6, 7].

———. *Masterspec,* Basic: Section 05810, Expansion Joint Cover Assemblies, 5/89 ed. The American Institute of Architects. [3, 4, 6, 7].

———. *Masterspec,* Basic: Section 06100, Rough Carpentry, 8/86 ed. The American Institute of Architects. [2].

———. *Masterspec,* Basic: Section 06170, Structural Glue Laminated Units, 5/88 ed. The American Institute of Architects. [2].

———. *Masterspec,* Basic: Section 06192, Prefabricated Wood Trusses, 8/86 ed. The American Institute of Architects. [2].

———. *Masterspec,* Basic: Section 06310, Heavy Timber Construction, 5/88 ed. The American Institute of Architects. [2].

———. *Masterspec,* Basic: Section 07600, Flashing and Sheet Metal, 5/88 ed. The American Institute of Architects. [3, 6, 7].

———. *Masterspec,* Basic: Section 07620, Metal Fasciae and Copings. 2/84 ed. The American Institute of Architects. [3, 4, 6, 7].

———. *Masterspec,* Basic: Section 07900, Joint Sealers, 8/87 ed. The American Institute of Architects. [7].

———. *Masterspec,* Basic: Section 09200, Lath and Plaster, 2/85 ed. The American Institute of Architects. [2].

———. *Masterspec,* Basic: Section 09215, Veneer Plaster, 2/88 ed. The American Institute of Architects. [2].

———. *Masterspec*, Basic: Section 09250, Gypsum Drywall, 8/87 ed. The American Institute of Architects. [2].

———. *Masterspec*, Basic: Section 09270, Gypsum Board Shaft Wall Systems, 2/88 ed. The American Institute of Architects. [2].

———. *Masterspec*, Basic: Section 09300, Tile, 2/86 ed. The American Institute of Architects. [2].

———. *Masterspec*, Basic: Section 09800, Special Coatings, 2/88 ed. The American Institute of Architects. [5, 6, 7].

———. *Masterspec*, Basic: Section 09900, Painting, 11/88 ed. The American Institute of Architects. [5, 6, 7].

———. *Masterspec*, Basic: Section 10100, Chalkboards and Tackboards, 11/85 ed. The American Institute of Architects. [3, 4, 6, 7].

———. *Masterspec*, Basic: Section 10160, Toilet Partitions, 5/87 ed. The American Institute of Architects. [3, 4, 5, 6, 7].

———. *Masterspec*, Basic: Section 10200, Louvers and Vents, 5/88 ed. The American Institute of Architects. [3, 4, 5, 6, 7].

———. *Masterspec*, Basic: Section 10440, Specialty Signs, 11/85 ed. The American Institute of Architects. [3, 4, 5, 6, 7].

———. *Masterspec*, Basic: Section 10500, Metal Lockers, 8/88 ed. The American Institute of Architects. [3, 4, 5, 6, 7].

———. *Masterspec*, Basic: Section 10522, Fire Extinguishers, Cabinets, and Accessories, 5/85 ed. The American Institute of Architects. [3, 4, 5, 6, 7].

———. *Masterspec*, Basic: Section 10605, Wire Mesh Partitions, 8/85 ed. The American Institute of Architects. [6, 7].

———. *Masterspec*, Basic: Section 10800, Toilet and Bath Accessories, 11/85 ed. The American Institute of Architects. [3, 4, 5, 6, 7].

———. *Masterspec*, Basic: Section 11050, Library Equipment, 3/87 ed. The American Institute of Architects. [3, 5, 6, 7].

———. *Masterspec*, Basic: Section 11400, Food Service Equipment, 8/85 ed. The American Institute of Architects. [3, 4, 5, 6, 7].

———. *Masterspec*, Basic: Section 11450, Residential Equipment, 11/82 ed. The American Institute of Architects. [3, 4, 5, 6, 7].

———. *Masterspec*, Basic: Section 11460, Unit Kitchens, 2/89 ed. The American Institute of Architects. [3, 4, 5, 6, 7].

Air Movement and Control Association. 1985. *Application Method for Air Louvers.* Arlington Heights, IL: Air Movement and Control Association. [6].

———. 1986. *Test Methods for Louvers, Dampers, and Shutters.* Arlington Heights, IL: Air Movement and Control Association. [6, 7].

Allen, Edward. 1985. *Fundamentals of Building Construction: Materials and Methods.* New York: Wiley. [3, 6].

American Architectural Manufacturers Association (AAMA). Aluminum Siding, Its Care and Maintenance (P-AS-1). Des Plaines, IL: AAMA. [6, 7].

————. Curtain Wall Manual #10 (CW-10). Des Plaines, IL: AAMA. [6, 7].

————. 1977. Voluntary Guide Specifications and Inspection Methods for Clear Anodic Finishes for Architectural Aluminum. (AAMA 607.1-1977). Des Plaines, IL: AAMA. [4].

————. 1977. Voluntary Guide Specifications and Inspection Methods for Electrolytically Deposited Color Anodic Finishes for Architectural Aluminum. (AAMA 608.1-1977). Des Plaines, IL: AAMA. [4].

————. 1976. Voluntary Guide Specifications and Inspection Methods for Integral Color Anodic Finishes for Architectural Aluminum. (AAMA 606.1-1976). Des Plaines, IL: AAMA. [4].

————. 1977. Voluntary Specifications for Residential Color Anodic Finishes. (AAMA 604.2-1977). Des Plaines, IL: AAMA. [4].

————. 1979. Voluntary Guide Specifications for Cleaning and Maintenance of Painted Aluminum Extrusions and Curtain Wall Panels. (AAMA 610.1-1979). Des Plaines, IL: AAMA. [5, 7].

————. 1985. Voluntary Guide Specifications for Cleaning and Maintenance of Architectural Anodized Aluminum. (AAMA 609.1-85). Des Plaines, IL: AAMA. [4, 7].

————. 1985. Voluntary Performance Requirements and Test Procedures for Pigmented Organic Coatings on Extruded Aluminum. (AAMA 603.8-85). Des Plaines, IL: AAMA. [5].

————. 1985. Voluntary Specifications for High Performance Organic Coatings on Architectural Extrusions and Panels. (AAMA 605.2-85). Des Plaines, IL: AAMA. [5].

————. 1986. Standard Specifications for Aluminum Siding, Soffits & Fascia. (AAMA 1402-86). Des Plaines, IL: AAMA. [6, 7].

American Concrete Institute. 1980. Guide for Concrete Floor and Slab Construction. (ACI 302.1R-80) American Concrete Institute. [2].

American Institute of Steel Construction (AISC). 1989. *Manual of Steel Construction*. Chicago, IL: AISC. [2, 6, 7]. Includes also the three following publications.

————. 1989. *Code of Standard Practice for Steel Buildings and Bridges*. Chicago, IL: AISC. [2, 6, 7]. Also contained in the AISC *Manual of Steel Construction*.

————. 1989. *Specifications for the Design, Fabrication, and Erection of Structural Steel for Buildings*. With supplemental vol., *Commentary*. Chicago, IL: AISC. [2, 6, 7]. Also contained in the AISC *Manual of Steel Construction*.

————. 1989. *Specifications for Structural Joints Using ASTM A 325 or A 490 Bolts*. Chicago, IL: AISC. [2, 6, 7]. Also contained in the AISC *Manual of Steel Construction*.

American Institute of Timber Construction. *Timber Construction Manual*. Englewood, CO: American Institute of Timber Construction. [2].

————. *Timber Construction Standards*. Englewood, CO: American Institute of Timber Construction. [2].

American Iron and Steel Institute. 1972. *Stainless Steel: Suggested Practices for Roofing, Flashing, Copings, Fascias, Gravel Stops, and Drainage*. Washington, DC: AISI. [3, 4, 6, 7].

————. 1975. Stainless Steel Fasteners: Suggested Source List. Washington, DC: AISI. [6, 7].

————. 1976. Stainless Steel Fasteners: A Systematic Approach to Their Selection. Washington, DC: AISI. [6, 7].

————. 1977. Design Guidelines for the Selection and Use of Stainless Steel. Washington, DC: AISI. [3, 4, 6, 7].

————. 1979. Welding of Stainless Steel and Other Joining Methods. Washington, DC: AISI. [6, 7].

American National Standards Institute (ANSI). Standard A14.3-1974. Safety Code for Fixed Ladders. ANSI. [6, 7].

————. Standard A58.1-1982. Minimum Design Loads, Buildings and Other Structures. ANSI. [2, 6, 7].

American Paint Contractor. 1988. Direct to Rust Coatings. *American Paint Contractor*, June, 65(6): 8–19. [3, 5].

American Society of Heating, Refrigerating and Air-Conditioning Engineers (ASHRAE). 1985. *ASHRAE Handbook; 1985 Fundamentals*. Atlanta, GA: ASHRAE. [6, 7].

American Society for Metals. 1982. *Metals Handbook: Ninth Edition, Vol. I: Properties and Selection of Metals*. Metals Park, OH: American Society for Metals. [3].

————. 1982. *Metals Handbook: Ninth Edition, Vol. II: Nondestructive Inspection and Quality Control*. Metals Park, OH: American Society for Metals. [2, 7].

————. 1982. *Metals Handbook: Ninth Edition, Vol. V: Surface Cleaning, Finishing, and Coating*. Metals Park, OH: American Society for Metals. [4, 5].

American Welding Society. 1983. *AWS D1.1 Structural Welding Code—Steel*. Miami, FL: AWS. [6, 7].

————. 1983. *AWS D1.2 Structural Welding Code—Aluminum*. Miami, FL: AWS. [6, 7].

American Wood Preservers Association. *Book of Standards*. Springfield, VA: American Wood Preservers Association. [2].

Appleman, Bernard R. 1989. Coatings for Steel Structures. *The Construction Specifier*, March, 42(3): 88–93. [5].

Architectural Graphic Standards. *See* Ramsey/Sleeper.

Architectural Technology. 1986. Technical Tips: Paints and Coatings Primer. *Architectural Technology*, July/August: 64–65. [5].

Ashurst, John. 1988. *Practical Building Conservation, Vol. 2*. Aldershot, England: Gower, **HP.**

Association for Preservation Technology. 1969. *Bulletins of the APT, Vol. 1*. Ottawa, Ont.: APT. **HP.**

ASTM. Standard A 6, Standard Specification for General Requirements for Rolled Steel Plates, Shapes, Sheet Piling, and Bars for Structural Use. ASTM. [2, 3, 6, 7].

——. Standard A 27, Standard Specification for Carbon-Steel Castings for General Application. ASTM. [3, 6].

——. Standard A 36, Standard Specification for Structural Steel. ASTM. [2, 3, 6].

——. Standard A 47, Specification for Ferritic Malleable Iron Castings. ASTM. [3, 6].

——. Standard A 48, Specification for Gray Iron Castings. ASTM. [3, 6, 7].

——. Standard A 53, Specification for Pipe, Steel, Black and Hot-Dipped, Zinc-Coated Welded and Seamless. ASTM. [3, 4, 6].

——. Standard A 108. Standard Specification for Steel Bars, Carbon, Cold-Finished, Standard Quality. ASTM. [3, 6].

——. Standard A 120, Specification for Pipe, Steel, Black and Hot-Dipped, Zinc-Coated (Galvanized) Welded and Seamless, for Ordinary Uses. ASTM. [3, 4, 6, 7].

——. Standard A 121, Specification for Zinc-Coated (Galvanized) Steel Barbed Wire. ASTM. [4, 6].

——. Standard A 123, Specification for Zinc (Hot-Galvanized) Coatings on Products Fabricated from Rolled, Pressed, and Forged Steel Shapes, Plates, Bars, and Strip. ASTM. [2, 3, 4, 6, 7].

——. Standard A 153, Specification for Zinc Coating (Hot-Dip) on Iron and Steel Hardware. ASTM. [3, 4, 6, 7].

——. Standard A 165, Standard Specification for Electrodeposited Coatings of Cadmium on Steel. ASTM [3, 4, 6, 7].

——. Standard A 167, Specification for Stainless and Heat-Resisting Chromium-Nickel Steel Plate, Sheet, and Strip, ASTM. [3, 4, 6, 7].

——. Standard A 269, Specification for Seamless and Welded Austenitic Stainless Steel Tubing for General Service. ASTM. [3, 4, 6, 7].

——. Standard A 276, Specification for Stainless and Heat-Resisting Steel Bars and Shapes. ASTM. [3, 4, 6, 7].

——. Standard A 283, Standard Specification for Low and Intermediate Tensile Strength Carbon Steel Plates, Shapes and Bars. ASTM. [3, 6, 7].

——. Standard A 307, Specification for Carbon Steel Externally Threaded Standard Fasteners. ASTM. [3, 6, 7].

——. Standard A 312, Specification for Seamless and Welded Austenitic Stainless Steel Pipe. ASTM. [3, 4, 6, 7].

——. Standard A 325, Specification for High-Strength Bolts for Structural Steel Joints. ASTM. [2, 3, 6, 7].

———. Standard A 366, Specification for Steel, Sheet, Carbon, Cold-Rolled, Commercial Quality. ASTM. [3, 6, 7].

———. Standard A 386, Specification for Zinc Coating (Hot-Dip) on Assembled Steel Products. ASTM. [3, 4, 6].

———. Standard A 392, Specification for Zinc-Coated Steel Chain-Link Fence Fabric. ASTM. [6, 7].

———. Standard A 424, Specification for Steel Sheet for Porcelain Enameling. ASTM. [3].

———. Standard A 446, Specifications for Sheet Steel, Zinc-Coated (Galvanized) by the Hot-Dip Process, Structural (Physical) Quality. ASTM. [3, 4, 6].

———. Standard A 463, Standard Specification for Sheet Steel, Cold-Rolled Aluminum-Coated Type 1. ASTM. [3, 4, 6].

———. Standard A 490, Specification for Heat-Treated Steel Structural Bolts, 150 ksi (1035 MPa) Tensile Strength. ASTM. [3, 6, 7].

———. Standard A 491, Specification for Aluminum-Coated Steel Chain-Link Fence Fabric. ASTM. [6, 7].

———. Standard A 500, Specification for Cold-Formed Welded and Seamless Carbon Steel Structural Tubing in Rounds and Shapes. ASTM. [3, 6].

———. Standard A 501, Specification for Hot-Formed Welded and Seamless Carbon Steel Structural Tubing. ASTM. [3, 6].

———. Standard A 510, Standard Specification for General Requirements for Wire Rods and Coarse Round Wire, Carbon Steel. ASTM. [3, 6].

———. Standard A 525, Specification for General Requirements for Steel Sheet Zinc-Coated (Galvanized) by the Hot-Dip Process. ASTM. [2, 3, 4, 6].

———. Standard A 526, Specification for Steel Sheet Zinc-Coated (Galvanized) by the Hot-Dip Process, Commercial Quality. ASTM. [3, 4, 6].

———. Standard A 527, Specifications for Steel Sheet Zinc-Coated (Galvanized) by the Hot-Dip Process, Lock-Forming Quality. ASTM. [3, 4, 6].

———. Standard A 528, Specification for Steel Sheet Zinc-Coated (Galvanized) by the Hot-Dip Process, Drawing Quality. ASTM. [3, 4, 6].

———. Standard A 536, Standard Specification for Ductile Iron Castings. ASTM. [3, 6, 7].

———. Standard A 554, Specification for Welded Stainless Steel Mechanical Tubing. ASTM. [3, 4, 6].

———. Standard A 568, Standard Specification for General Requirements for Steel, Carbon and High-Strength Low-Alloy Hot-Rolled Sheet and Cold-Rolled Sheet. ASTM. [3, 6].

———. Standard A 569, Specification for Steel, Carbon (0.15 Maximum, Percent), Hot-Rolled Sheet and Strip, Commercial Quality. ASTM. [3, 6]

———. Standard A 570, Specification for Hot-Rolled Carbon Steel Sheet and Strip, Structural Quality. ASTM. [3, 6].

———. Standard A 572, Specification for High-Strength Low-Alloy Columbium-Vanadium Steels of Structural Quality. ASTM. [3, 6, 7].

———. Standard A 575, Standard Specification for Steel Bars, Carbon, Merchant Quality, M-Grades. ASTM. [3, 6, 7].

———. Standard A 585, Specification for Aluminum-Coated Steel Barbed Wire. ASTM. [6, 7].

———. Standard A 591, Specification for Steel Sheet, Cold-Rolled Electrolytic Zinc-Coated. ASTM. [3, 4, 6].

———. Standard A 606, Specification for Steel Sheet and Strip, Hot-Rolled, High-Strength, Low-Alloy, with Improved Atmospheric Resistance. ASTM. [3].

———. Standard A 607, Standard Specification for Steel Sheet and Strip, Hot-Rolled and Cold-Rolled, High-Strength, Low-Alloy Columbium and or Vanadium. ASTM. [3].

———. Standard A 611, Specification for Steel, Cold-Rolled Sheet, Carbon, Structural. ASTM. [3, 6, 7].

———. Standard A 641, Specification for Zinc-Coated (Galvanized) Carbon Steel Wire. ASTM. [2, 3, 4, 6, 7].

———. Standard A 642, Standard Specification for Steel Sheet, Zinc-Coated (Galvanized) by the Hot-Dip Process, Drawing Quality, Special Killed. ASTM. [3, 4].

———. Standard A 663, Standard Specification for Steel Bars, Carbon, Merchant Quality, Mechanical Qualities. ASTM. [3, 6, 7].

———. Standard A 675, Standard Specification for Steel Bars, Carbon, Hot-Wrought, Special Quality, Mechanical Properties. ASTM. [3, 6, 7].

———. Standard A 743, Standard Specification for Castings, Iron-Chromium-Nickel, and Nickel-Base, Corrosion-Resistant, for General Application. ASTM. [3, 6, 7].

———. Standard A 744, Standard Specification for Castings, Iron-Chromium-Nickel, and Nickel-Base, Corrosion-Resistant, for Severe Service. ASTM. [3, 4, 6, 7].

———. Standard A 780, Practice for Repair of Damaged Hot-Dip Galvanized Coatings. ASTM. [4].

———. Standard A 786, Standard Specification for Rolled Steel Floor Plates, ASTM. [3, 6].

———. Standard A 792, Standard Specification for Steel Sheet, Aluminum–Zinc Alloy-Coated by the Hot-Dip Process. ASTM. [4].

———. Standard A 875, Standard Specification for Steel Sheet, Zinc 5 Percent, Aluminum-Mische Metal Alloy, Coated by the Hot-Dip Process. ASTM. [4, 6, 7].

———. Standard B 26, Specification for Aluminum-Alloy Sand Castings. ASTM. [3, 6, 7].

———. Standard B 32, Specification for Solder Metal. ASTM. [6, 7].

———. Standard B 43, Specification for Seamless Red Brass Pipe, Standard Sizes. ASTM. [3, 4, 6].

———. Standard B 62, Specification for Composition Bronze or Ounce Metal Castings. ASTM. [3, 4, 6, 7].

———. Standard B 86, Specification for Zinc-Alloy Die Castings. ASTM. [6].

———. Standard B 101, Standard Specification for Lead-Coated Copper Sheets. ASTM. [4, 6].

———. Standard B 108, Specification for Aluminum-Alloy Permanent Mold Castings. ASTM. [3, 6, 7].

———. Standard B 134, Specification for Brass Wire. ASTM. [3, 6].

———. Standard B 135, Specification for Seamless Brass Tube. ASTM. [3, 4, 6].

———. Standard B 177, Recommended Practice for Chromium Electroplating on Steel for Engineering Use. ASTM. [3, 4, 6].

———. Standard B 209, Specification for Aluminum and Aluminum Alloy Sheet and Plate. ASTM. [3, 6, 7].

———. Standard B 210, Specification for Aluminum-Alloy Drawn Seamless Tubes. ASTM. [3, 6, 7]

———. Standard B 211, Specification for Aluminum-Alloy Bars, Rods, and Wire. ASTM. [3, 6, 7].

———. Standard B 221, Specification for Aluminum-Alloy Extruded Bars, Rods, Wire, Shapes, and Tubes. ASTM. [3, 6, 7].

———. Standard B 247, Specification for Aluminum-Alloy Die and Hand Forgings. ASTM. [3, 6, 7]

———. Standard B 253, Recommended Practice for Preparation of and Electroplating on Aluminum Alloys by the Zincate Process. ASTM. [3, 4].

———. Standard B 254, Recommended Practice for Preparation and Electroplating on Stainless Steel. ASTM. [3,4].

———. Standard B 271, Specification for Copper-Base Alloy Centrifugal Castings. ASTM. [3, 4].

———. Standard B 320, Recommended Practice for Preparation of Iron Castings for Electroplating. ASTM. [3, 4].

———. Standard B 370, Specification for Copper Sheet and Strip for Building Construction. ASTM. [4, 6, 7].

———. Standard B 429, Specification for Aluminum-Alloy Extruded Structural Pipe and Tube. ASTM. [3, 6, 7].

———. Standard B 455, Specification for Copper-Zinc-Lead Alloy (Leaded Brass) Extruded Shapes. ASTM. [3, 4, 6, 7].

———. Standard B 456, Specification for Electrodeposited Coatings of Copper Plus Nickel Plus Chromium and Nickel Plus Chromium. ASTM. [3, 4, 6, 7].

———. Standard B 483, Specification for Aluminum and Aluminum-Alloy Extruded Drawn Tubes for General Purpose Applications. ASTM. [3, 6, 7].

————. Standard B 584, Specification for Copper Alloy Sand Castings. ASTM. [3, 4, 6, 7].

————. Standard B 650, Specification for Electrodeposited Engineering Chromium Coatings on Ferrous Substrates. ASTM. [3, 4, 6, 7].

————. Standard C 282, Test Method for Acid Resistance of Porcelain Enamels (Citric Acid Spot Test). ASTM. [4].

————. Standard C 283, Test Method for Resistance of Porcelain Enamel Utensils to Boiling Acid. ASTM. [4].

————. Standard C 286, Definition of Terms Relating to Porcelain Enamel and Ceramic-Metal Systems. ASTM. [4].

————. Standard C 313, Test Method for Adherence of Porcelain Enamel and Ceramic Coatings to Sheet Metal. ASTM. [4].

————. Standard C 346, Test Method for 45-deg Specular Gloss of Ceramic Materials. ASTM. [4].

————. Standard C 448, Test Method for Abrasion Resistance of Porcelain Enamel. ASTM. [4].

————. Standard C 481, Test Method for Laboratory Aging of Sandwich Construction. ASTM. [3, 6, 7]

————. Standard C 538, Test Method for Color Retention of Red, Orange, and Yellow Porcelain Enamels. ASTM. [4].

————. Standard C 540, Test Method for Image Gloss of Porcelain Enamel Surfaces. ASTM. [4].

————. Standard C 578, Specification for Preformed Cellular Polystyrene Thermal Insulation. ASTM. [6].

————. Standard C 641, Test Method for Staining Materials in Lightweight Concrete Aggregates. ASTM. [2].

————. Standard C 645, Specification for Non-Load (Axial) Bearing Steel Studs, Runners (Track), and Rigid Furring Channels for Screw Application of Gypsum Board. ASTM. [2].

————. Standard C 646, Specification for Steel Drill Screws for Application of Gypsum Board to Light-Gage Steel Studs. ASTM. [2].

————. Standard C 665, Specification for Mineral Fiber Blanket Thermal Insulation for Light Frame Construction and Manufactured Housing. ASTM. [3, 6].

————. Standard C 703, Test Method for Spalling Resistance of Porcelain Enameled Aluminum. ASTM. [4].

————. Standard C 754, Specification for Installation of Steel Framing Members to Receive Screw-Attached Gypsum Wallboard, Backing Board, or Water-Resistant Backing Board. ASTM. [2].

————. Standard C 840, Specification for Application and Finishing of Gypsum Board. ASTM. [2].

————. Standard C 841, Specification for Installation of Interior Lathing and Furring. ASTM. [2].

————. Standard C 919, Practices for Use of Sealants in Acoustical Applications. ASTM. [6, 7].

————. Standard C 920, Specification for Elastomeric Joint Sealants. ASTM. [6, 7].

————. Standard C 955, Specification for Load-Bearing (Transverse and Axial) Steel Studs, Runners (Track), and Bracing or Bridging, for Screw Application of Gypsum Board and Metal Plaster Bases. ASTM. [2].

————. Standard C 1007, Installation of Load-Bearing (Transverse and Axial) Steel Studs and Accessories. ASTM. [2].

————. Standard C 1063, Specification for Installation of Lathing and Furring for Portland Cement–Based Plaster. ASTM. [2].

————. Standard D 16, Standard Definitions of Terms Relating to Paint, Varnish, Lacquer, and Related Products. ASTM. [5].

————. Standard D 226, Specification for Asphalt and Asphalt-Saturated Organic Felt Used in Roofing and Waterproofing. ASTM. [2, 6, 7].

————. Standard D 659, Evaluating Degree of Chalking of Exterior Paints. ASTM. [5].

————. Standard D 660, Evaluating Degree of Checking of Exterior Paints. ASTM. [5].

————. Standard D 661, Evaluating Degree of Cracking of Exterior Paints. ASTM. [5].

————. Standard D 662, Evaluating Degree of Erosion of Exterior Paints. ASTM. [5].

————. Standard D 1187, Test Method for Asphalt Emulsions for Use as Protective Coatings for Metal. ASTM. [3, 4, 5, 7].

————. Standard D 1653, Test Method for Moisture Vapor Permeability of Organic Coating Films. ASTM. [5].

————. Standard D 1730, Recommended Practices for Preparation of Aluminum and Aluminum-Alloy Surfaces for Painting. ASTM. [4, 5].

————. Standard D 1731, Recommended Practices for Preparation of Hot-Dip Aluminum Surfaces for Painting. ASTM. [4, 5].

————. Standard D 1779, Specification for Adhesive for Acoustical Materials. ASTM. [6, 7].

————. Standard D 2092, Recommended Practices for Preparation of Zinc-Coated Steel Surfaces for Painting. ASTM. [4, 5].

————. Standard D 2833, Standard Index of Methods for Testing Architectural Paints and Coatings. ASTM. [5].

————. Standard D 3276, Standard Guide for Painting Inspectors (Metal Substrates). ASTM. [5].

————. Standard D 3794, Practice for Testing Coil Coatings. ASTM. [4, 5].

————. Standard D 3927, Standard Guide for State and Institutional Purchasing of Paint. ASTM. [5].

———. Standard E 84, Test Method for Surface Burning Characteristics of Building Materials. ASTM. [2, 5, 6, 7].

———. Standard E 94, Recommended Practice for Radiographic Testing. ASTM. [2, 7].

———. Standard E 96, Test Method for Water Vapor Transmission of Materials. ASTM. [5].

———. Standard E 97, Test Method for 45-deg, 0-deg Directional Reflectance Factor of Opaque Specimens by Broad-Band Filter Reflectometry. ASTM. [4].

———. Standard E 136, Test Method for Behavior of Materials in a Vertical Tube Furnace at 750 degrees C. ASTM. [5].

———. Standard E 142, Controlling Quality of Radiographic Testing. ASTM. [2].

———. Standard E 164, Recommended Practice for Ultrasonic Contact Examination of Weldments. ASTM. [2, 6, 7].

———. Standard E 165, Recommended Practice for Liquid Penetrant Inspection. ASTM. [2].

———. Standard E 709, Practice for Magnetic Particle Examination. ASTM. [2].

———. Standard E 894, Test Method for Anchorage of Permanent Metal Railing Systems and Rails for Buildings. ASTM. [6, 7].

———. Standard E 935, Test Method for Performance of Permanent Metal Railing Systems and Rails for Building Systems. ASTM. [6, 7].

———. Standard E 985, Specification for Permanent Metal Railing Systems and Rails for Buildings. ASTM. [6, 7].

———. Standard F 552, Definition of Terms Relating to Chain Link Fencing. ASTM. [6, 7].

———. Standard F 567, Practice for Installation of Chain-Link Fence. ASTM. [7].

———. Standard F 626, Specification for Fence Fittings. ASTM. [6].

———. Standard 668, Specification for Poly(Vinyl Chloride) (PVC)-Coated Steel Chain-Link Fence Fabric. ASTM. [6].

Bakhalov, G. T. and A. V. Turkovakaya. 1965. *Corrosion and Protection of Metals.* Elmsford, NY: Pergamon Press. [3, 4, 5, 7].

Banov, Able. 1973. *Paints and Coatings Handbook for Contractors, Architects and Builders.* Farmington, MI: Structures Publishing Co. [5].

Bartlett, Thomas L. 1983. Cleaning with Corncobs. *The Construction Specifier,* Feb., 36(2): 6–7. [5].

Batcheler, Penelope Hartshorne. 1968. Paint Color Research and Restoration. Technical Leaflet 15. Nashville, TN: American Association for State and Local History. **HP.** [5].

Bauer, Mike. 1987. Letterbox: Too Much to the Imagination? *The Construction Specifier,* July, 40(7): 7–8. [5].

Bennett, C. R. 1987. Paints and Coatings: Getting Beneath the Surface. *The Construction Specifier,* Feb., 40(2): 36–41. [5].

Black, David R. Dealing with Peeling Paint. *North Carolina Preservation*, Dec. 1986–Feb. 1987, 66: 16–17. [5].

Bocchi, Greg. 1988. Powder Coatings Making Inroads in Metal Construction Market. *Metal Architecture*, April, 4(4): 8–9. [5].

Bower, Norman F. 1985. Insurance by the Gallon. *The Construction Specifier*, April, 38(4): 96–99. [5].

British Iron and Steel Research Association (BISRA). 1977. *How to Prevent Rusting*. London: BISRA. [4, 5, 7].

Brunnell, Gene. 1977. *Built to Last: A Handbook on Recycling Old Buildings*. Washington, DC: Preservation Press. **HP.**

Building Design and Construction. 1988. Focus on Metals in Building Construction. *Building Design and Construction*, June, 29(6): 67. [1, 2, 3, 4, 5, 6, 7].

Canadian Heritage. 1985. Take It All Off? Advice on When and How to Strip Interior Paintwork. *Canadian Heritage*, Dec.–Jan.: 44–47. [5].

Catani, Mario J. 1985. Protection of Embedded Steel in Masonry. *The Construction Specifier*, Jan., 38(1): 62–68. [2, 3, 4, 5, 7].

Chain Link Fence Manufacturers Institute (CLFMI). 1989. *Product Manual*. Washington, DC: CLFMI. [6].

Chase, Sarah B. 1984. Home Work: The ABC's of House Painting. *Historic Preservation*, Aug., 36(4): 12–14. [5].

City Limits. 1986. Exterior Paints. *City Limits*, Oct., 11(8): 15. [5].

Commerce Publishing Corp. 1988. *The Woodbook*. Seattle, WA: Commerce Publishing Corp. [2].

Commercial Remodeling. 1980. Technique: Steel Floor Ideas. *Commercial Remodeling*, Oct., 2(5): 26. [6, 7].

———. 1981. Technique: Ion Painting. *Commercial Remodeling*. Feb., 3(1): 24. [5].

Commercial Renovation. 1984. Technique: Remodeling America's Number-One Symbol Is a Long, Slow Process. *Commercial Renovation*, Dec., 6(6): 23–24. [4].

———. 1984. Technique: Curved Aluminum Canopies Offer New Look for Renovated Buildings. *Commercial Renovation*, Aug., 6(4): 20. [6, 7].

———. 1987. The 1987 Premier Renovation Architects. *Commercial Renovation*, Dec., 9(6): 24–44. **HP.**

Construction Specifications Institute (CSI). 1988. CSI Monograph Series 07M411, Precoated Metal Building Panels. Alexandria, VA: Construction Specifications Institute. [4, 5].

———. 1988. *Specguide* 09900, Painting, Alexandria, VA: Construction Specifications Institute. [5].

Construction Specifier, The. 1985. Painting to Protect. *The Construction Specifier*, April, 38(4): 92–94, 123. [5].

Copper Development Association. Building Expansion Joints. Copper Development Association. [6, 7].

————. Clear Film-Coated Copper for Decorative Applications. Copper Development Association. [5].

————. Clear Organic Finishes for Copper and Copper Alloys. New York: Copper Development Association. [5].

————. *Copper, Brass, Bronze Design Handbook, Architectural Applications.* New York: Copper Development Association. [3, 4, 6, 7].

————. 1980. *Copper, Brass, Bronze Design Handbook, Sheet Copper Applications.* New York: Copper Development Association. [3, 4, 6, 7].

————. Design Guide: Forgings. New York: Copper Development Association. [3, 4, 6, 7].

————. How to Apply Statuary and Patina Finishes. New York: Copper Development Association. [4].

————. Joining Copper–Nickel Alloys. Copper Development Association. [6].

————. Joining Copper–Tin Alloys (Phosphor Bronzes). Copper Development Association. [6].

————. Mechanical Properties of Soldered Copper. Copper Development Association. [6].

————. Properties of Clear Organic Coatings on Copper and Copper Alloys. New York: Copper Development Association. [5].

————. Sheet Copper Fundamentals. Copper Development Association. [3, 4, 6, 7].

————. Soldering and Brazing Copper Tube. Copper Development Association. [6].

————. Standard Designations for Copper and Copper Alloys. New York: Copper Development Association. [3, 4, 6, 7].

————. *Standards Handbook, Part 2—Alloy Data.* New York: Copper Development Association. [3, 4, 6, 7].

————. *Standards Handbook, Part 3—Terminology.* New York: Copper Development Association. [3, 4, 6, 7].

————. *Standards Handbook, Part 4—Engineering Data.* New York: Copper Development Association. [3, 4, 6, 7].

————. *Standards Handbook, Part 7—Cast Products.* New York: Copper Development Association. [3, 4, 6, 7].

————. Taking Care of the Metal in Your Building. Copper Development Association. [4, 7].

————. *Welding Handbook.* New York: Copper Development Association. [6, 7].

D'Angelo, Charles and Owen J. Perryman. 1988. Insulated Panels and the Pre-Engineered Building Market. *Metal Architecture,* Oct. 4(10): 20. [6, 7].

Dean, Sheldon W. and T. S. Lee, eds. 1988. *Degradation of Metals in the Atmosphere.* Philadelphia, PA: ASTM. [3].

Evans, U. R. 1972. *The Corrosion and Oxidation of Metals: Scientific Principles and Practical Applications.* London: Edward Arnold, Ltd. [3].

———. 1972. *The Rusting of Iron: Causes and Control, Studies in Chemistry No. 7*. London: Edward Arnold, Ltd. [3].

Federal Specifications. *See* U.S. General Services Administration Specifications Unit.

Fisher, Thomas. 1983. When the Rains Come. *Architecture*, July. [5].

Fishman, Herbert B. 1986. Architectural Copper Work. *The Construction Specifier*, Nov., 39(11): 60–66. [3, 4, 6].

Forest Products Laboratory. *Handbook No. 72—Wood Handbook*. Washington, DC: U.S. Department of Agriculture. [2].

Gregerson, John. 1988. Metal Panels Have Designs on New Markets. *Building Design and Construction*, June, 29(6): 68–71. [2, 3, 4, 5, 6, 7].

Gypsum Association. 1985. *Recommended Specifications for Application and Finishing of Gypsum Board*. (GA-216-85). Evanston, IL: Gypsum Association. [2].

———. 1985. *Using Gypsum Board for Walls and Ceilings*. (GA-201-85). Evanston, IL: Gypsum Association. [2].

———. 1986. *Recommendations for Covering Existing Interior Walls and Ceilings with Gypsum Board*. (GA-650-86). Evanston, IL: Gypsum Association. [2].

———. 1986. *Recommended Specifications: Gypsum Board Types, Uses, Sizes and Standards*. (GA-223-86). Evanston, IL: Gypsum Association. [2].

Hardingham, David. 1980. Preparing for Painting. *The Old-House Journal*, Oct.: 133–36 [5].

Harsfield, S. W. 1982. A Valid Question. *The Construction Specifier*, Dec., 35(12): 7.

Harvey, John. 1972. *Conservation of Buildings*. London: Baker. **HP.**

Historic Preservation. 1983. Home Work. *Historic Preservation*, July/Aug., 35(4): 10–11. **HP.**

———. 1984. Home Work: Spring Fixes. *Historic Preservation*, April, 36(2): 14–17. **HP.**

Hornbostel, Caleb. 1978. *Construction Materials, Types, Uses, and Applications*. New York: Wiley.

Howell, J. Scott. 1987. Architectural Cast Iron: Design and Restoration. *The Construction Specifier*, July, 40(7): 70–74.

Insall, Donald W. 1972. *The Care of Old Buildings Today: A Practical Guide*. London: Architectural Press. **HP.**

Jentsch, Bryan C. 1988. Pre-Insulated Panels Providing Excellent Economic Value. *Metal Architecture*, Oct., 4(10): 21. [6].

Johnson, Stephen. 1988. Improvement in Domestic Steel Quality Translating into Better Metal Paneling. *Metal Architecture*, Sept., 4(9): 5, 85. [3, 4, 5, 6, 7].

Jones, Larry. 1984. Painting Galvanized Metal. *The Old-House Journal*, Jan.–Feb., 12(1): 10–11. [4, 5].

Kincaid, Mary. 1982. What Paint Experts Say. *The Construction Specifier*, July, 35(5): 54–61. [5].

Koller, Alice. 1981. Hot-Dip Galvanizing: How and When to Use It. *The Construction Specifier*, Sept., 34(8): 47–51. [4, 6].

LaQue, F. E. and H. R. Copson, eds. 1963. *Corrosion Resistance of Metals and Alloys*, 2nd ed. New York: Van Nostrand Reinhold. [3].

Lead Industries Association. *Lead Roofing and Flashing*. New York: Lead Industries Association. [3].

Leidheiser, Henry, Jr. 1971. *The Corrosion of Copper, Tin, and Their Alloys*. New York: Wiley. [3].

Lowes, Robert. 1988. Abrasive Blasting. *PWC Magazine*, May/June, 50(3): 26–30. [4, 5].

Mahowald, Dave. 1988. Specifying Paint Coatings for Harsh Environments. *The Construction Specifier*, Oct. 41(10): 13–16. [5].

Martens, Charles R., Sherwin-Williams Company. 1974. *Technology of Paints and Lacquers*. New York: Robert E. Krieger Publishing Co. [5].

Maruca, Mary. 1984. 10 Most Common Restoration Blunders. *Historic Preservation*, Oct. 36(5): 13–17. **HP.**

McDonald, Timothy B. 1987. Technical Tips: Coatings That Protect Against the Corrosion of Steel. *Architecture*, July: 101–102. [5].

———. 1989. Specifying Welding Details. *Architecture*, Jan.: 113–14. [6, 7].

Metal Architecture. 1987. Building Repainting System Duplicates Coil Coatings in Appearance, Life Expectancy. *Metal Architecture*, Nov., 3(11): 24. [5].

———. 1988. 1988 Guide to Architectural Coil Coatings. *Metal Architecture*, Nov. 4(11): 22–23. [4, 5].

Metal Construction News. 1988. *1989 Building Systems Product File & Directory*. Skokie, IL: Metal Construction News. [6, 7].

Metal Lath/Steel Framing Association. 1985. *Lightweight Steel Framing Systems Manual*, 2nd ed. Chicago, IL: Metal Lath/Steel Framing Association. [2].

———. 1986. *Specifications for Metal Lathing and Furring*. Chicago, IL: Metal Lath/Steel Framing Association. [2].

———. *Technical Bulletin No. 18: Fire Rated Metal Lath/Steel Stud Exterior*. Chicago, IL: Metal Lath/Steel Framing Association. [2].

———. *Technical Bulletin No. 101: Types of Metal Lath and Their Uses*. Chicago, IL: Metal Lath/Steel Framing Association. [2].

———. *Technical Bulletin No. 120: Ceramic Tile Applied to Metal Lath and Plaster*. Chicago, IL: Metal Lath/Steel Framing Association. [2].

———. 1986. *Technical Information File*. Chicago, IL: Metal Lath/Steel Framing Association. [2].

Modern Metals. 1989. Finishing Forum: Coil Anodized Sheet Survives with Flying Colors. *Modern Metals*, Feb., 45(1): 22–28. [4].

———. 1989. Presses on Roll Form Lines: How to Choose the Best. *Modern Metals*, Feb., 45(1): 10–20. [4, 5].

Moit, Dan. 1988. Coatings for Metals: Preplan the Selection. *Metal Architecture*, Sept., 4(9): 8. [4, 5].

Monnich, Joni. 1983. Restorer's Notebook: Beware Brass-Cleaner Damage. *The Old-House Journal*, Oct., 11(8): 180. [4].

Moormann, Ambrose F., Jr. 1982. Paint and the Prudent Specifier. *The Construction Specifier*, July, 35(5): 69–71. [5].

Moreno, Elena Marcheso. 1987. Failures Short of Collapse. *Architecture*, July: 91–94. [2, 7].

Munger, Charles G. 1984. *Corrosion Prevention by Protective Coatings*. Houston, TX: National Association of Corrosion Engineers. [5].

National Association of Architectural Metal Manufacturers (NAAMM). 1979. *Metal Bar Grating Manual*. Chicago, IL: NAAMM. [6, 7].

———. 1979. *Welding Standard for Fabrication of Metal Bar Gratings*. NAAMM. [6, 7].

———. 1982. *Metal Stairs Manual*. Chicago, IL: NAAMM. [6, 7].

———. 1986. *Heavy Duty Metal Bar Grating Manual*. Chicago, IL: NAAMM. [6, 7].

———. 1986. *Pipe Railing Manual*. Chicago, IL: NAAMM. [6, 7].

———. 1988. *Metal Finishes Manual for Architectural and Metal Products*. Chicago, IL: NAAMM. [4, 5].

National Decorating Products Association. 1988. *Paint Problem Solver*. St. Louis, MO: National Decorating Products Association. Also available from Painting and Decorating Contractors of America. [5].

National Fire Protection Association. 1988. *NFPA 101: Life Safety Code*. Quincy, MA: NFPA. [6, 7].

National Forest Products Association. *Manual for House Framing*. Washington, DC: National Forest Products Association. [2].

———. *National Design Specifications for Wood Construction*. Washington, DC: National Forest Products Association. [2].

———. *Span Tables for Joists and Rafters*. Washington, DC: National Forest Products Association. [2].

National Trust for Historic Preservation. 1985. *All About Old Buildings—The Whole Preservation Catalog*. Washington, DC: Preservation Press. An extensive reference containing names and addresses of many organizations active in historic preservation, and lists of publications' sources. Anyone facing a preservation problem should obtain this catalog, which will save much time in finding the right organization or data source. **HP.**

O'Donnell, Bill. 1988. Refinishing Bathroom Fixtures. *The Old-House Journal*, Nov./Dec., 16(6): 22–25. [4]. **HP.**

Old-House Journal, The. 1982. Stripping Paint. *The Old-House Journal*, Dec.: 249–52. [5]. **HP.**

———. 1983. Ask OHJ: The Gilt Complex. *The Old-House Journal*, March, 11(2): 49. [4]. **HP.**

———. 1983. Ask OHJ: Paint on Paint. *The Old-House Journal*, Nov., 11(9): 202. [5]. **HP.**

———. 1983. 48 Paint Stripping Tips. *The Old-House Journal*, March, 11(2): 44–45. [5]. **HP.**

———. 1983. Our Opinion of 'Peel Away.' *The Old-House Journal*, May, 11(4): 80. [5]. **HP.**

———. 1985. Stripping Clinic. *The Old-House Journal*, Dec., 13(10): 212B. [5]. **HP.**

———. 1987. Exterior Painting: Problems and Solutions. *The Old-House Journal*, Sept./Oct.: 35–39. [5]. **HP.**

———. 1988. Commercial Paint Stipping. *The Old-House Journal*, July/Aug., 16(4): 29–33. [5]. **HP.**

Olin, Harold B.; John L. Schmidt, and Walter H. Lewis. 1983. *Construction Principles, Materials and Methods*, 5th ed. Chicago, IL: U.S. League of Savings Institutions. [3, 4, 5, 6, 7].

Osburn, Donald H. and John M. Foehl. 1963. Coloring and Finishing of Copper Metals. *The Construction Specifier*, Oct., 16(5): 50–55. [4].

Painting and Decorating Contractors of America. 1975. *Painting and Decorating Craftsman's Manual and Textbook*, 5th ed. Falls Church, VA: Painting and Decorating Contractors of America. [5].

———. 1982. *Painting and Decorating Encyclopedia*. Falls Church, VA: Painting and Decorating Contractors of America. [5].

———. 1984. *Painting and Wallcovering: A Century of Excellence*. Falls Church, VA: Painting and Decorating Contractors of America. [5].

———. 1986. *Architectural Specification Manual, Painting, Repainting, Wallcovering and Gypsum Wallboard Finishing*, 3rd ed. Kent, WA: Specifications Services, Washington State Council of PDCA. [5].

———. 1988. *Hazardous Waste Handbook*. Falls Church, VA: Painting and Decorating Contractors of America. [5].

———. 1988. *The Master Painters Glossary*. Falls Church, VA: Painting and Decorating Contractors of America. [5].

Petersen, Maurice R. 1984. Finishes on Metals: A View from the Field—Part 1. *The Construction Specifier*, 37(12)(Dec.): 36–39. [4, 5].

———. 1985. Finishes on Metals: A View from the Field—Part 2. *The Construction Specifier*, 38(1)(Jan.): 70–73. [4, 5].

Phillips, Morgan W. 1975. Some Notes on Paint Research and Reproduction. Technical Bulletin 7(4): 14–19. Ottawa: Association for Preservation Technology. [5].

Porcelain Enamel Institute (PEI). Bulletin T-2, Test for Resistance of Porcelain Enamel to Abrasion. PEI. [4].

———. Bulletin T-20, Image Gloss Test. PEI. [4].

———. Bulletin T-21, Test for Acid Resistance of Porcelain Enamels. PEI. [4].

———. Bulletin T-22, Cupric Sulfate Test for Color Retention. PEI. [4].

———. Bulletin T-51, Antimony Trichloride Spall Test for Porcelain Enameled Aluminum. PEI. [4].

———. 1969. ALS-105(69) Recommended Specification for Architectural Porcelain Enamel on Aluminum for Exterior Use. PEI. [4, 6].

———. 1970. *Guide to Designing with Architectural Porcelain Enamel on Steel.* PEI. [4, 6].

———. 1970. S-100(65). Specification for Architectural Porcelain Enamel on Steel for Exterior Use. PEI. [4, 6].

———. 1986. Color Guide for Architectural Porcelain Enamel. PEI. [4].

———. 1973. The Weatherability of Porcelain Enamel. PEI. [4].

Powell, Bill. 1988. Patina in Paradise. *Metal Architecture,* Aug., 4(8): 20–21. [4].

Powers, Alice Leccese. 1984. Young Master of the Gilders Art. *Historic Preservation,* Dec., 36(6): 43–46. [4].

Ramsey/Sleeper and the AIA Committee on Architectural Graphic Standards. 1981. *Architectural Graphic Standards,* 7th ed. New York: Wiley. [2, 3, 4, 5, 6, 7].

Reader's Digest Association. 1973. *Reader's Digest Complete Do-It-Yourself Manual.* Pleasantville, NY: The Reader's Digest Association. [3, 5, 6, 7].

Reynolds, Patrick T. 1984. The Threat to Outdoor Art. *Historic Preservation,* June, 36(3): 35–39. [4].

Rich, Jack C. 1947. Reprinted 1988. *The Materials and Methods of Sculpture.* New York: Dover. [3, 4, 5, 6, 7].

Robertson, Stanley, 1989. Restoring the Gilded Surface. *The Construction Specifier,* 42(7)(July): 69–76. [4].

Scharfe, Thomas R. 1988. New Metal Coating Technologies Enhance Design Opportunities. *Building Design and Construction,* June, 29(6): 86–89. [5].

Scott, Gerald. 1965. *Atmospheric Oxidation and Antioxidants.* New York: Elsevier. [3].

Sealant Engineering and Associated Lines (SEAL). 1987. SEAL Guide Specification for Sealants and Caulking. San Diego, CA: SEAL. [6, 7].

Sheet Metal and Air Conditioning Contractors National Association (SMACNA). 1987. *Architectural Sheet Metal Manual.* Vienna, VA: SMACNA. [6, 7].

Sherwin-Williams Co. 1988. Painting and Coating Systems for Specifiers and Applicators. Sherwin-Williams. [5].

Sivinski, Valerie A. 1986. Preserving Historic Materials: Ferrous Metals. *Architecture,* Nov.: 108–109. **HP.** [3, 5].

Southern Pine Inspection Bureau. *Standard Grading Rules for Southern Pine Lumber.* Pensacola, FL: Southern Pine Inspection Bureau. [2].

Southworth, Susan and Michael. 1978. *Ornamental Ironwork: An Illustrated Guide to Its Design, History, and Use in American Architecture.* Boston: David R. Godine. [3, 6, 7].

Staehli, Alfred M. 1985. Historic Preservation: Where to Find the Facts. *The Construction Specifier,* July, 38(7): 50–53. **HP.**

Stahl, Frederick A. 1984. *A Guide to the Maintenance, Repair, and Alteration of Historic Buildings.* New York: Van Nostrand Reinhold. **HP.**

Stanwood, Les. 1983. Acid Rain: A Cloudy Issue. *The Construction Specifier,* Nov., 36(11): 74–79. [3, 4, 5].

Steel Structures Painting Council. 1983. *Steel Structures Painting Manual, Vol. 1: Good Paint Practice,* 2nd ed. Pittsburgh, PA: Steel Structures Painting Council. [5].

———. 1983. *Steel Structures Painting Manual, Vol. 2: Systems and Specifications,* 2nd ed. Pittsburgh, PA: Steel Structures Painting Council. [5].

Stern, E. George. 1986. Pre-Standardization Research Project on Reliability of Permanent Railing Systems and Rails for Buildings. *Fabricator,* Sept.–Oct.: 23. [4, 6].

———. 1987. Stiffness vs. Strength. *Fabricator,* Jan.–Feb.: 36–39. [4, 6].

Stoddard, Brooke C. 1987. Home Work: Picking the Right Paint Color. *Historic Preservation,* Sept.–Oct., 39(5): 16–19. **HP.**

Tatum, Rita. 1986. Metal Building Systems Go High-Rise High-Tech. *The Construction Specifier,* March, 39(3): 40–49. [6, 7].

———. 1986. Restoring America's Beacon of Freedom. *The Construction Specifier,* July, 39(7): 34–44. [4, 5].

Ting, Raymond. 1986. Metal Panel Behavior in Exterior Wall Design. *Exteriors.* Autumn: 65–69. [6, 7].

Truss Plate Institute. *Design Specifications for Light Metal Plate Connected Wood Trusses.* Madison, WI: Truss Plate Institute. [2].

Underwriters Laboratories. *Building Materials Directory—Class A, B, C: Fire and Wind Related Deck Assemblies.* Northbrook, IL: Underwriters Laboratories. [2].

———. *Fire Resistance Directory—Time/Temperature Constructions.* Northbrook, IL: Underwriters Laboratories. [2].

United States Gypsum Company. 1972. *Red Book: Lathing and Plastering Handbook,* 28th ed. Chicago, IL: United States Gypsum Company. [2].

———. 1987. *Gypsum Construction Handbook,* 3rd ed. Chicago, IL: United States Gypsum Company. [2].

U.S. Department of the Army. *Technical Manual TM 5-801-2. Historic Preservation Maintenance Procedures.* Washington, DC: Department of the Army. **HP.**

———. 1980. *Painting: New Construction and Maintenance,* EM 1110-2-3400. Washington, DC: Superintendent of Documents, U.S. Government Printing Office. [5].

————. 1983. CRD-C621—Specification for Nonshrink Grout. Department of the Army. [7].

————. 1982. Corps of Engineers Guide Specifications, Military Construction, Section 05120, Structural Steel. Office of the Chief of Engineers, Department of the Army. [2, 3, 6, 7].

————. 1982. Corps of Engineers Guide Specifications, Military Construction, Section 05141, Welding, Structural. Office of the Chief of Engineers, Department of the Army. [2, 6, 7].

————. 1981. Corps of Engineers Guide Specifications, Military Construction, CEGS-05500, Miscellaneous Metals. Office of the Chief of Engineers, Department of the Army. [3, 4, 5, 6, 7].

————. 1980. Corps of Engineers Guide Specifications, Military Construction, Section 07600, Sheet Metalwork, General. Office of the Chief of Engineers, Department of the Army. [3, 4, 5, 6, 7].

————. 1978. Corps of Engineers Guide Specifications, Military Construction, CE-246.01. Partitions, Movable; Flush, Semiflush, and Panel Types. Office of the Chief of Engineers, Department of the Army. [6, 7].

————. 1982. Corps of Engineers Guide Specifications, Military Construction, CE-240.01, Furring (Metal), Lathing, and Plastering. Office of the Chief of Engineers, Department of the Army. [2].

————. 1981. Corps of Engineers Guide Specifications, Military Construction, CEGS-09910, Painting, General. Office of the Chief of Engineers, Department of the Army. [5].

————. 1980. Corps of Engineers Guide Specifications, Military Construction, CEGS-11701. Casework, Metal and Wood. Office of the Chief of Engineers, Department of the Army. [6, 7].

U.S. Departments of the Army and the Air Force. 1965. *Building Construction Materials and Practices: Caulking and Sealing.* Washington, DC: U.S. Government Printing Office. [6, 7].

U.S. Departments of Army, the Navy, and the Air Force. 1969. *Technical Manual TM 5-618, Paints and Protective Coatings.* Washington, DC: U.S. Government Printing Office. [5].

U.S. Department of Commerce. *PS 1—U.S. Product Standard for Construction and Industrial Plywood.* Washington, DC: U.S. Department of Commerce. [2].

————. *PS 20—American Softwood Lumber Standard.* Washington, DC: U.S. Department of Commerce. [2].

————. 1968. *Organic Coatings BSS 7.* Washington, DC: U.S. Department of Commerce, National Bureau of Standards. [5].

U.S. Department of Defense. 1984. Military Specification MIL-C-285808(YD). Cabinet, Storage; Wardrobe, Three Drawer. DOD. [6, 7].

————. 1970. Military Specification MIL-P-21035. Paint, High Zinc Dust Content, Galvanizing Repair. DOD. [4, 5].

————. 1977. Military Specification MIL-W-28581A(YD), Wardrobes: Clothing, Composite Wood and Steel. DOD. [6, 7].

U.S. Department of the Navy, Naval Facilities Engineering Command. 1983 (Oct.). Guide Specifications Section 02444. Fence, Chain Link. Department of the Navy. [6].

————. 1985 (April). Guide Specifications Section 05120. Structural Steel Department of the Navy. [2, 3, 6, 7].

————. 1985 (Oct.). Guide Specifications Section 05500. Metal Fabrications. Department of the Navy. [3, 4, 5, 6, 7].

————. 1983 (March). Guide Specifications Section 06100. Rough Carpentry. Department of the Navy. [2].

————. 1985 (Oct.). Guide Specifications Section 07600. Flashing and Sheet Metal. Department of the Navy. [3, 6, 7].

————. 1981 (Aug.). Guide Specifications Section 07920. Sealants and Caulkings. Department of the Navy. [6, 7].

————. 1981 (May). Guide Specifications Section 08330. Coiling Steel Doors. Department of the Navy. [6, 7].

————. 1984 (Feb.). Guide Specifications Section 09100. Metal Support Systems. Department of the Navy. [6, 7].

————. 1986 (Feb.). Guide Specifications Section 09910C. Painting of Buildings (Field Painting). Department of the Navy. [5].

————. 1982 (Nov.). Guide Specifications Section 10162. Toilet Partitions, Department of the Navy. [6, 7].

————. 1983 (Jan.). Guide Specifications Section 10201. Metal Wall and Door Louvers. Department of the Navy. [6, 7].

————. 1985 (Oct.). Guide Specifications Section 10440. Signs. Department of the Navy. [6, 7].

————. 1983 (Oct.). Guide Specifications Section 10800. Toilet and Bath Accessories. Department of the Navy. [6, 7].

————. 1982 (Aug.). Guide Specifications Section 12322. Wardrobes. Department of the Navy. [6, 7].

————. 1983 (Oct.). Guide Specifications Section 13121. Pre-engineered Metal Buildings. Department of the Navy. [2, 6, 7].

U.S. General Services Administration. 1977 (Oct.). Public Building Services Guide Specification, Section 4-07600. Sheet Metal, Flashing and Related Accessories. U.S. General Services Administration. [3, 6, 7].

————. 1971 (Feb.). Public Building Services Guide Specification, Section 3-0990. Painting and Finishing, Renovation, Repair and Improvement. U.S. General Services Administration. [5].

————. 1973 (Oct.). Public Building Services Guide Specification, Section 4-0990.01. Painting and Finishing. U.S. General Services Administration. [5].

U.S. General Services Administration Specifications Unit. Federal Specifications.

————. AA-L-00486H(1). Lockers, Clothing, Steel. Washington, DC: GSA Specifications Unit. [6, 7].

————. FF-B-561. Bolts (Screw), Lag. Washington, DC: GSA Specifications Unit. [6, 7].

————. FF-B-575C. Bolts, Hexagon and Square. Washington, DC: GSA Specifications Unit. [6, 7].

————. FF-B-588. Bolt, Toggle, and Expansion Sleeve, Screw. Washington, DC: GSA Specifications Unit. [6, 7].

————. FF-S-92. Screw, Machine, Slotted, Cross Recessed or Hexagon Head. Washington, DC: GSA Specifications Unit. [6, 7].

————. FF-S-111. Screw, Wood. Washington, DC: GSA Specifications Unit. [6, 7].

————. FF-S-325. Shield, Expansion; Nail, Expansion; and Nail, Drive Screw (Devices, Anchoring, Masonry). Washington, DC: GSA Specifications Unit. [6, 7].

————. FF-W-84. Washers, Lock (Spring). Washington, DC: GSA Specifications Unit. [6, 7].

————. FF-W-92. Washers, Flat (Plain). Washington, DC: GSA Specifications Unit. [6, 7].

————. HH-I-521F. Insulation Blankets, Thermal (Mineral Fiber, for Ambient Temperatures). Washington, DC: GSA Specifications Unit. [6, 7].

————. HH-I-558B. Insulation, Blocks, Boards, Blankets, Felts, Sleeving (Pipe and Tube Covering), and Pipe Fitting Covering, Thermal (Mineral Fiber, Industrial Type). Washington, DC: GSA Specifications Unit. [6, 7].

————. QQ-F-461C. Floor Plate, Steel, Rolled. Washington, DC: GSA Specifications Unit. [3, 6].

————. QQ-L-101. Lath, Metal, (and Other Metal and Plaster Bases). Washington, DC: GSA Specifications Unit. [6].

————. RR-F-191H/GEN. Fencing, Wire and Post Metal (and Gates, Chain Link Fence Fabric, and Accessories) (General Specification). Washington, DC: GSA Specifications Unit. [6].

————. RR-F-191/1B. Fencing, Wire and Post Metal (Chain Link Fence Fabric) (Detail Specification). Washington, DC: GSA Specifications Unit. [6].

————. RR-F-191/2B. Fencing, Wire and Post Metal (Chain Link Fence Gates) (Detail Specification). Washington, DC: GSA Specifications Unit. [6].

————. RR-F-191/3B. Fencing, Wire and Post Metal (Chain Link Fence Posts, Top Rails and Braces) (Detailed Specification). Washington, DC: GSA Specifications Unit. [6].

————. RR-F-191/4B. Fencing, Wire and Post Metal (Chain Link Fence Accessories) (Detailed Specification). Washington, DC: Specifications Unit. [6].

————. RR-T-650. Treads, Metallic and Non-Metallic. Washington, DC: GSA Specifications Unit. [6].

————. TT-C-535. Coating, Epoxy, Two-Component, For Interior Use on Metal, Wood, Wallboard, Painted Surfaces, Concrete, and Masonry. Washington, DC: GSA Specifications Unit. [5].

————. TT-C-542. Coating, Polyurethane, Oil-free, Moisture Curing. Washington, DC: GSA Specifications Unit. [5].

————. TT-E-489G. Enamel, Alkyd, Gloss (For Exterior and Interior Surfaces). Washington, DC: GSA Specifications Unit. [5].

————. TT-E-496. Enamel, Heat Resisting, (400 deg F), Black. Washington, DC: GSA Specifications Unit. [5].

————. TT-E-505A. Enamel, Odorless, Alkyd, Interior, High Gloss, White and Light Tints. Washington, DC: GSA Specifications Unit. [5].

————. TT-E-508C. Enamel, Interior, Semigloss, Tints and White. Washington, DC: GSA Specifications Unit. [5].

————. TT-E-509B. Enamel, Odorless, Alkyd, Interior, Semigloss, White and Tints. Washington, DC: GSA Specifications Unit. [5].

————. TT-E-545. Enamel, Odorless, Alkyd, Interior, Undercoat, Flat, Tints and White. Washington, DC: GSA Specifications Unit. [5].

————. TT-E-1593. Enamel, Silicone Alkyd Copolymer, Gloss (for Exterior and Interior Use). Washington, DC: GSA Specifications Unit. [5].

————. TT-L-190D. Linseed Oil, Boiled (For Use in Organic Coatings). Washington, DC: GSA Specifications Unit. [5].

————. TT-L-201A. Linseed Oil, Heat Polymerized. Washington, DC: GSA Specifications Unit. [5].

————. TT-P-28F. Paint, Aluminum, Heat Resisting (1,200 F.). Washington, DC: GSA Specifications Unit. [5].

————. TT-P-30E. Paint, Alkyd, Odorless, Interior, Flat White and Tints. Washington, DC: GSA Specifications Unit. [5].

————. TT-P-37D. Paint, Alkyd Resin; Exterior Trim, Deep Colors. Washington, DC: GSA Specifications Unit. [5].

————. TT-P-47F. Paint, Oil, Nonpenetrating-flat, Ready-mixed Tints and White (for Interior Use). Shakes and Rough Siding. Washington, DC: GSA Specifications Unit. [5].

————. TT-P-81E. Paint, Oil, Alkyd, Ready Mixed Exterior, Medium Shades. Washington, DC: GSA Specifications Unit. [5].

————. TT-P-86G. Paint, Red-lead-base, Ready-mixed. Washington, DC: GSA Specifications Unit. [5].

————. TT-P-615D. Primer Coating, Basic Lead Chromate, Ready Mixed. Washington, DC: GSA Specifications Unit. [5].

————. TT-P-636D. Primer Coating, Alkyd, Wood and Ferrous Metal. Washington, DC: GSA Specifications Unit. [5].

———. TT-P-641G. Primer Coating; Zinc Dust–Zinc Oxide (For Galvanized Surfaces). Washington, DC: GSA Specifications Unit. [5].

———. TT-P-645A. Primer, Paint, Zinc Chromate, Alkyd Type. Washington, DC: GSA Specifications Unit. [5].

———. TT-P-664C. Primer Coating, Synthetic, Rust-inhibiting, Lacquer-resisting. Washington, DC: GSA Specifications Unit. [5].

———. TT-P-1511A. Paint, Latex-base, Gloss and Semi-gloss, Tints and White (for Interior Use). [5].

———. TT-T-291F. Thinner, Paint, Mineral Spirits, Regular and Odorless. [5].

———. TT-V-86C. Varnish, Oil, Rubbing (For Metal and Wood Furniture). [5].

Waite, John G. 1986. Caring for Decorative Metalwork. *Historic Preservation,* Nov.–Dec., 38(6): 20–24. [3, 4, 5, 6, 7].

Weaver, Martin E. 1989. Fighting Rust, Part I: A Backgrounder. *The Construction Specifier,* 42(5)(May): 143–45. [3, 5].

———. 1989. Fighting Rust, Part II: Remedies. *The Construction Specifier,* 42(6)(June): 129–30. [5].

———. 1989. Caring for Bronze. *The Construction Specifier,* 42(7)(July): 58–66. [4, 5].

Weinstein, Nat. 1984. How to Match Paint Colors. *The Old-House Journal,* Jan.–Feb., 12(1): 7–9. [5]. **HP.**

Weismantel, Guy E., ed. 1981. *Paint Handbook.* New York: McGraw-Hill. [5].

West Coast Lumber Inspection Bureau. *Standard Grading Rules for West Coast Lumber.* Portland, OR: West Coast Lumber Inspection Bureau. [2].

Western Lath, Plaster and Drywall Contractors' Association. 1988. *Plaster/Metal Framing System/Lath Manual.* New York: McGraw-Hll. [2].

Western Wood Products Association (WWPA). *Grading Rules for Western Lumber.* Portland, OR: WWPA. [2].

———. *Grade Stamp Manual.* Portland, OR: WWPA. [2].

———. *A-2, Lumber Specifications Information.* Portland, OR: WWPA. [2].

———. *Western Woods Use Book.* Portland, OR: WWPA. [2].

———. *Wood Frame Design.* Portland, OR: WWPA. [2].

Wilson, Forrest. 1984. *Building Materials Evaluation Handbook.* New York: Van Nostrand Reinhold. [2–7].

———. 1985. Building Diagnostics. *Architectural Technology,* Winter: 22–41. [2, 3].

Wright, Gordon, 1987. Trends in Specifying Architectural Coatings. *Building Design and Construction,* June, 28(6): 188–92. [5].

Zinc Institute. *Zinc Coatings for Corrosion Protection.* New York: Zinc Institute. May not be available since the Zinc Institute is no longer in operation. [4].

Index

Substrates *(cont.)*
  immediate failure, 176, 177, 178
  solid, 59
  solid problems, 60–61
  supported, 58
Support systems, 22–62
  concrete and steel structural
    framing systems, 23–26
  information sources, 61–62
  installing new metal framing and
    furring over existing materials,
    41
  installing new wood framing and
    furring over existing materials,
    58
  and metal fabrications, 226
  metal framing and furring, 26–41
  miscellaneous building elements, 61
  substrates, 58–61
  wood framing and furring, 41–58

Telegraphing, 185, 197, 213
Temper designations, aluminum, 80–
  81
Terneplate, 128, 214
Thin metals
  dents in, 290
  welding of, 252
Thinners, 162–63
Tin, 91
Tinplate, 128
Toilet accessories, 235
Tool steel, 70
Toxic wastes, 210
Truss-type floor joists, 45

Unified Numbering System (UNS),
  76, 86
Unit kitchens, 233
Urethane, 162

Vapor degreasing, 107
Vinyl chloride copolymers, 162
Vitreous coatings, 133–35
Volatile solvents, 162–63

Wall caps, 237
Weathering and exposure, 93
Welding, 237, 240, 251–53
White cast iron, 68
White rust, 125–26
Window-cleaning devices, 230
Window stools, 236
Wire mesh, 239
Wood framing and furring, 42–58
  over existing materials, 58
  failures, 52–58
  general requirements, 42–43
  incorrectly constructed, 57–58
  installation, 47–52
  misaligned, warped, or twisted, 57
  miscellaneous framing, 51–52
  movement, 53–55
  problems, 52–55
  repairing and extending, 55–58
  shrinkage, 57
  spacings, 49–50
  stud walls and partitions, 50–51
  wood framing definition, 42
  wood and plywood, 43–47
Wood and plywood
  for framing and furring, 43–48
  installation, 47–52
  lumber classification and grades,
    43–44
  lumber sizes, 44
  lumber species, 44
  materials, 43–46
  moisture content of lumber, 43
  roof trusses, 45
  softwood plywood, 44
  treatment, 46–47
  truss-type floor joists, 45
Wrinkling, of coatings, 181, 185, 186,
  187, 191, 203–4
Wrought alloys, aluminum, 79–80
Wrought iron, 65–67

Zinc, 89–90, 121, 124
Zinc-alloy metals, 26, 240
Zincating, 129, 141
Zinc Institute, 101, 153
Zinc-plated aluminum, 129